CAMBRIDGE LIBRARY COLLECTION

Books of enduring scholarly value

Botany and Horticulture

Until the nineteenth century, the investigation of natural phenomena, plants and animals was considered either the preserve of elite scholars or a pastime for the leisured upper classes. As increasing academic rigour and systematisation was brought to the study of 'natural history', its subdisciplines were adopted into university curricula, and learned societies (such as the Royal Horticultural Society, founded in 1804) were established to support research in these areas. A related development was strong enthusiasm for exotic garden plants, which resulted in plant collecting expeditions to every corner of the globe, sometimes with tragic consequences. This series includes accounts of some of those expeditions, detailed reference works on the flora of different regions, and practical advice for amateur and professional gardeners.

The Gentlewoman's Book of Gardening

The 'Victoria Library for Gentlewomen', a series of books 'Under the Patronage of H.M. the Queen and H.R.H. the Princess of Wales', edited by W. H. Davenport Adams (1828–91), provided information and advice on various topics for those who aspired to gentlewomanly status. Davenport Adams himself was a journalist and author of popular science and history works, but little is known of the two authors of this 1892 work. Edith L. Chamberlain was a minor novelist who had also published a book on the dialect of west Worcestershire, and Fanny Douglas worked with Davenport Adams on other titles in the series. This book follows the fashion of late nineteenth–century works (often by women) which combine descriptions of gardens and gardening with historical and literary references. It is unusual in that its final chapter describes ways for educated 'gentlewomen' to enter gardening as a profession – a radical suggestion for the period.

Cambridge University Press has long been a pioneer in the reissuing of out-of-print titles from its own backlist, producing digital reprints of books that are still sought after by scholars and students but could not be reprinted economically using traditional technology. The Cambridge Library Collection extends this activity to a wider range of books which are still of importance to researchers and professionals, either for the source material they contain, or as landmarks in the history of their academic discipline.

Drawing from the world-renowned collections in the Cambridge University Library and other partner libraries, and guided by the advice of experts in each subject area, Cambridge University Press is using state-of-the-art scanning machines in its own Printing House to capture the content of each book selected for inclusion. The files are processed to give a consistently clear, crisp image, and the books finished to the high quality standard for which the Press is recognised around the world. The latest print-on-demand technology ensures that the books will remain available indefinitely, and that orders for single or multiple copies can quickly be supplied.

The Cambridge Library Collection brings back to life books of enduring scholarly value (including out-of-copyright works originally issued by other publishers) across a wide range of disciplines in the humanities and social sciences and in science and technology.

The Gentlewoman's Book of Gardening

EDITH L. CHAMBERLAIN
FANNY DOUGLAS

CAMBRIDGE
UNIVERSITY PRESS

CAMBRIDGE
UNIVERSITY PRESS

University Printing House, Cambridge, CB2 8BS, United Kingdom

Cambridge University Press is part of the University of Cambridge.
It furthers the University's mission by disseminating knowledge in the pursuit of
education, learning and research at the highest international levels of excellence.

www.cambridge.org
Information on this title: www.cambridge.org/9781108076623

© in this compilation Cambridge University Press 2017

This edition first published 1892
This digitally printed version 2017

ISBN 978-1-108-07662-3 Paperback

Under the Patronage of H.M. the Queen and
H.R.H. the Princess of Wales

THE VICTORIA LIBRARY FOR GENTLEWOMEN

THE

GENTLEWOMAN'S

BOOK OF GARDENING

EDITED BY W. H. DAVENPORT ADAMS

Yrs faithfully
G. L. Chamberlain

THE

GENTLEWOMAN'S
BOOK OF GARDENING

BY
EDITH L. CHAMBERLAIN
AND
FANNY DOUGLAS

" Flowers only flourish rightly in the garden of one who loves them
RUSKIN

LONDON:
HENRY AND CO.,
BOUVERIE STREET, E.C.
1892

Printed by Hazell, Watson, & Viney, Ld., London and Aylesbury

CONTENTS.

———◆———

THE GENTLEWOMAN'S BOOK OF GARDENING.

———◆———

CHAPTER I.

THE GARDEN IN ROMANCE.

THE garden has ever been the headquarters of romance. It is at once the type of the Hebrew Paradise and of the Greek Elysium. On the one hand we have Milton, who sings—

> "Not that fair field
> Of Enna, where Prosérpine gathering flowers,
> Herself a fairer flower, by gloomy Dīs
> Was gather'd, which cost Ceres all that pain
> To seek her through the world; nor that sweet grove
> Of Daphne by Orontes, and th' inspired
> Castalian spring, might with this Paradise
> Of Eden strive."

Nothing in literature has surpassed the poet's description of the virginal pair enshrined therein.

1

On the other hand, Homer—a veritable chronicler
of primeval joys—has chosen the garden as the
setting of some of his tenderest scenes.   What
garden is there to equal that of Nausicaa?

> " There in full prime the orchard trees grow tall—
> Sweet fig, pomegranate, apple, fruited fair,
> Pear and the healthful olive."

Less pleasing in its associations, but equally delightful in its details, was that of Calypso, where

> " All about a meadowy ground was seen
> Of violets mingling with the parsley green."

And there is no tenderer and more touching
scene in all the grand old poem than that in
which Ulysses, after long years of absence, finds
his old father labouring humbly in his garden,
and gently chides him with caring more for its
adornment than his own.

The Thracian kings had rose-gardens on the
sheltered slopes of the snowy Bernicus ; and
the garden of the Hesperides, with the dragon-
guarded fruit, was on the northern coast of Africa.
The hanging gardens of Babylon (one of the
wonders of the world) were constructed for a
home-sick Median queen by her Assyrian lover.
Now, where once these were, there blooms only

one strange and solitary tree, brought ages since from some alien shore.

Tasso tells us of "Armida's enchanted realm," and Schiller relates how "Die schönen Tage in Aranjuez sind nun zu ende." It was in Aranjuez that, for the delight of Queen Isabella, the desert was forced to blossom as the rose, and the barren soil of Castile made splendid with flowers.

Two pretty and pathetic legends of garden-queens we may recall. At Tacubaya, in Mexico, exists still, beside a clump of trees, the "Fountain of the Queen." Once there was a lovely garden round the fountain, and daily Queen Malinche and her maidens repaired thither to bathe. One day, just after they had laid aside their white robes, a party of Aztec hunters surprised them. Malinche did not hesitate. Rather than endure their gaze she plunged headlong into the fountain, never to rise again. At noon, however, "people of acute vision" see her gold and vermilion head-dress floating there, and know that for a moment she has risen from her crystal home below to gaze again on the garden she once loved.

The other is a pretty story of a beautiful Moorish maiden named Galiana, whose father built for her a "lordly pleasure house," surrounded by delicious gardens made quaint with fantastic kiosks and cool with marble baths and splashing fountains. But amidst all its beauty Galiana was unhappy, for a great and gruff Moorish giant would persist in wooing her. But one day a gallant young Frankish prince came riding along, saw and loved the maiden, slew the too importunate giant, and won Galiana for his own.

Many a garden has been laid out to give pleasure to beauty. Gabrielle d'Estrées had a noteworthy garden, and Fair Rosamond was imprisoned in another. Marie Antoinette lived in the golden age of gardens, and had the Petit Trianon, with its dairy, farms, chapel, and cottage for the curé, reconstructed to please herself. Here, clad in dainty white, with be-witching gauze fichus and flower-wreathed straw hats, she and her ladies played like children on the edge of the volcano. They fished in the lake, superintended the milking of the cows, and tried to make butter in the dairy.

Almost as pathetic is the child-garden of Mary Queen of Scots, whose traces may still be seen in the placid islet in Lake Menteith. Here, at least, her memory is pure and beautiful and innocent, for here she lived only as a child.

The gentlewoman is associated with the garden in fancy more than in reality. It was in a garden that the ill-fated lovers of Verona met and loved; it was in a garden that Paolo and Francesca learned the love they immortalised; it was in the garden of the King of Navarre that certain merry lords and ladies we wot of scorned, derided, and conquered each other. There was also the garden tended "from morn to even" by

" A Lady, the wonder of her kind,
Whose form was upborne by a lovely mind,
Which, dilating, had moulded her mien and motion
Like a sea-flower unfolded beneath the ocean."

And here it was that the " sensitive plant" and other lovely and ethereal flowers grew. Linked closely with the garden, too, is the thought of "simple Isabel," and the poor remains of her lover, hid within a garden-pot, with a plant of basil growing over them. Rose, the gardener's daughter, lived within a garden; roses grew over its porch, lilac perfumed the air, whilst a privet hedge sheltered it from the road, and a green wicket gate admitted Eustace and his friend.

It was in Windsor garden that James I. of Scotland got the first glimpse of his lady love, "the fairest and the freshest younge flower of all." "Ah sweete!" he says,

> "'are ye a worldly creáture,
> Or heavenly thing in likeness of natúre?

> "'Or are ye god Cupidé's own princéss,
> And comen are to loose me out of band?
> Or are ye very Nature, the goddéss
> That have depainted with your heavenly hand
> This garden full of flowers as they stand?'"

As lovely as any was the garden of Schalimas,—

> "With foot as light
> As the young musk-roe's out she flew,
> To cull each shining leaf that grew
> Beneath the moonlight's hallowing beams,
> For this enchanted wreath of dreams—
> Anemones and seas of gold,
> And new-born lilies of the river,
> And those sweet flowerets that unfold
> Their buds on Camadena's quiver."

Nor must we forget Saccharissa's walk in the garden of Penshurst. Penshurst was Sidney's home, and there stands

> "The taller tree, which of a nut was set
> At his great birth, when all the Muses met."

Saccharissa was, of course, Lady Dorothy Sidney, the inspirer of much of Waller's verse.

It is only of late years that women have

assumed their right place as the presiding genii
of the garden, and have even attempted pro-
fessionally to grow and care for flowers, to
arrange and look after window-boxes, to design
home and table decorations. They could not
well find a more agreeable pursuit, for surely
all true women will agree that the fragrant air
of the garden is sweeter than the dim and dusty
atmosphere of the lecture-room, that the cult
of flowers is more befitting and more enjoyable
than the frenzied pursuit of a vote. Gardening
soothes and calms the mind, whilst public strife
unsettles the temper and destroys tranquillity.
Women, like the "great Diocletian," should know
when they are well off. Cowley says—

> "Methinks I see great Diocletian walk
>   In the Salonian garden's noble shade,
>   Which by his own imperial hands was made.
> I see him smile, methinks, as he does talk
>   With the ambassadors, who come in vain
>   To entice him to a throne again.
> 'If I, my friends,' said he, 'should to you show
>   All the delights which in these gardens grow,
>   'Tis likelier much that you with me would stay,
>   Than 'tis that you should carry me away.'"

There are still many ways in which a gentlewoman might occupy herself in a garden more than
she does at present. It is a pity that the pleasant

old institution, the still-room, has been allowed to fall into disuse. In the old days the country dame was her own chemist and apothecary. She grew her own medicinal and culinary herbs, and distilled or dried them, as the case might be. Why should this charming and useful pastime not be revived? Chemistry is a common item in education nowadays, and here is a way in which the girl-scientist may apply her new knowledge for the good of her household. The object of the enlightened woman of to-day should be not to discard household and domestic duties, but to apply to them the results of recent researches in hygiene, economics, science, and art, so as to elevate what was once mere domestic drudgery to an honoured and honourable occupation. In a later chapter we hope to show how the still-room may be advantageously and scientifically revived.

Another pleasant old use of the garden was the cultivation of " strewings." As everybody knows, carpets are a modern innovation ; and, by comparison, " strewings " are apt to seem to us barbaric and unattractive. But under good housewives the " strewings " (which seem to have included wall decoration) were carefully chosen. At Christmas rosemary, bay, mistletoe, and holly were used.

> " Rosemary and rose—these keep
> Seeming, and savour all the winter long,"

says Shakespeare ; and at Christmas, when the
customary reunion reawakens the sense of loss, what
more fitting than that rosemary, the garniture of
graves, the type of remembrance, should strew the
floors and scent the rooms ?

At Candlemas these were replaced by box and
Easter yew ; and at Whitsuntide they, in turn,
gave way to the freshly budded twigs of birch.
In summer rushes and cool oaken boughs were
considered the best coverture.

Many another fragrant herb was used as well,
however, and Drayton gives a comprehensive list
of these : —

> " Sweet Lavender—with Rosemary and Bays ;
> Sweet Marjoram, with her like, sweet Basil, rare for smell ;
> The heathful Balm and Mint,—
> The scentful Camomile, the vertuous Costmarie ;
> Clear Hyssop, and therewith the comfortable Thyme ;
> Germander with the rest, each thing then in her prime :
> Among these strewing kinds some others wild that grow,
> As Burnet, all abroad, and Meadowwort they throw."

Herrick also details at length the succession of
plants used for strewing the floors and walls.

It was about the beginning of the sixteenth
century that house-decoration by plants first began,

oleanders, myrtles, etc., being then placed about the rooms.

It has been suggested by one writer that a Shakespearean garden might be laid out, where the devotee of the " swan of Avon " might muse at will concerning the flowers and their associations. There would be pansies and rue for Ophelia ; pleached honeysuckle for Beatrice, and all Perdita's bouquet—daffodils, violets, primroses, oxslips; with love-in-idleness for Oberon and Titania ; red and white roses, like those that grew in the Temple gardens, and fantastic box trees, such as they affected in Illyria. This idea might be carried further ; amaranths, daffodils, and " cowslips wan " might once more unite their tears in memory of Lycid, and

> " Violets blue,
> And fresh-blown roses washed in dew,"

proclaim the sovereignty of " mirth and youthful jollity." In some favoured and sheltered spot—if necessary, under glass—the companions of the sensitive plant might again range themselves around her ; and in another, roses, lilies, and larkspur might once more look and long for Maud.

The change in the spirit of gardening within

the last two hundred years must not be overlooked. Long ago no garden was complete without statuary and playing fountains; but now these are no longer essentials. The unromantic conservatory has replaced the stony dryads and broken-nosed nymphs of old. Sun-dials are growing scarce, and peacocks seem less plentiful than heretofore. Trees are no longer cut into fantastic devices, but have been liberated, like other slaves, from the bondage of former days.

Strange, indeed, must have been some of the ancient gardens. In Pliny's garden the box trees were cut into the shape of animals, and others, again, imitated pyramids and similar formal devices. In Elizabethan gardens the flower-beds were divided by gravel walks, and edged with box, thrift, and thyme.

The taste for curious conceits in everything was characteristic of Elizabethan days. "Literary millinery," as Morley calls it, was in its zenith, and the same strange taste showed itself in gardening. Giles Fletcher describes a particularly singular design—

> "The garden like a lady fair was cut,
> That lay as if she slumber'd in delight,
> And to the open skies her eyes did shut.
>   *   *   *   *

" Upon a holly bank her head she cast,
    On which the bower of vain delight was built ;
    White and red roses for her face were placed,
    And for her tresses marigolds were spilt."

Never since then has this sort of thing been carried
to such excess. Milton's Eden opened up a vista
of new possibilities to gardeners, and Pope and
Addison completed the downthrow of excessive
formality.

The "notes" of last century gardening and of
Le Notre's masterpieces were spaciousness and
elegance—formality so vastly conceived, that it
ceased to be offensive. Heraldic devices wrought
in flower-beds were common in these gardens,
however. Chantilly and the Tuileries were
carpeted with the arms of France.

In Holland there was at one time more extra-
ordinary fantasy in trees than anywhere else. A
row of fine elms was cut to simulate a stag hunt ;
and in Flanders yews were allowed to degrade
themselves into uncouth caricatures of geese and
turkeys.

But apart from all associations of poetry, legend,
or romance, surely a garden is a delight in itself.
It has had no more faithful and fervent lover
than Gerarde,—the Isaac Walton of the garden.

Flowers, he says, are designed " to adorne the garlands of the Muses, to decke the bosoms of the beautiful, to paint the gardens of the curious, to garnish the glorious crownes of kings." And again he asks—

" Whither did the poets hunt for their syncere delights but into the gardens of Alcinous, of Adonis, and the orchards of Hesperides ? Where did they dream that heaven should be but in the pleasant garden of Elysium ? " And Gerarde, when reproached for discursiveness and inaccuracy, excused himself on the ground that he " wrote for gentlewomen,"—as we do, though, indeed, we protest that we have striven to be truly accurate.

Of famous gardens it needs not to speak. Marly, Versailles, Chantilly, Stowe, Moor Park, and Twickenham, Nonsuch, Windsor, and Chatsworth, are all well known. The floating gardens of the Mexicans, the sacred herb gardens (odoriferous of garlic) of the Egyptians, the Provençal Paradise of King René, are among those famed in history.

In recent literature two pleasant pictures stand out before all others. One is that of the garden of Hornby Mills, as described by Henry Kingsley. It was tended only by two old gentlewomen and a cripple boy. In January the quaint-cut beds were all outlined in gold and green by thick-set golden aconite. Scarce had these gallant little pioneers begun to droop before a line of dainty hepaticas was ready to take their place. Crocuses followed, and all the year there was a pleasant plenitude of flowers, the whole described with a grace and a vivacity that fix it in the mind for ever. The other is the garden of Ham House, as described by Miss Mitford. She tells of " its deep silent courts," and says " there were no kalmias, azaleas, or magnolias," but only " flowers of the olden time." Her charming account makes the garden likeable, although we see no reason why " flowers of the olden time " and the splendid American invaders should not grow peacefully and happily together.

Mahomet pictured the Paradise of the true believers as a garden with the Tuba tree of happiness growing in the centre of it. The tree may not be visible at first sight, but it grows in the middle of every garden, and sooner or later is found by each true lover of flowers.

CHAPTER II.

GARDEN PARAPHERNALIA.

MANY, in these days, are the trials of the amateur. On the one hand, professors of all Arts, Crafts, and Sciences are eager for pupils, and spare no pains to infuse into their scholars so much knowledge that ambition is aroused. The amateurs enter the lists with the professionals, who now, on the other hand, begin to grumble about over-competition, and view with jealousy and covert sneers the best efforts of their ex-pupils. Then there are the manufacturers, who strew our path with descriptions of pianos that will all but play themselves; genuine Cremonas to be had for an old song ; cheap paint-boxes that look so attractive ; futile appliances in gimcrack caskets; sets of miniature tools, pretty, but worthless.

The horticultural implement-maker may not sin to the same extent as some of his compeers, but this is chiefly for want of power, not of will. He cannot pack forks and trowels in velvet-lined boxes of (paper) morocco, putting all the value into the case, and subtracting it from the tools. But he does, nevertheless, set traps for unwary feet, especially feminine ones. The moral of this, fair gardeners, runs thus: Be not persuaded into buying little toy sets of "ladies'" tools. Rather choose each implement separately; select it of dimensions that you can wield comfortably; though it need not, because it is small, of necessity be weak-backed and insecure, of untempered metal or unseasoned wood. Assuming (as we do) that our readers garden because it suits their tastes to do so, the following list of tools will suffice. If well-chosen and judiciously used, work may be done with them of appreciable value in the garden economy. A trowel must, of course, head the number, not that it is of such overweening importance, but because it is distinctively the amateur's tool. Many, pinning their whole faith to it, fancy themselves gardeners immediately they set up a trowel, and never advance to the use of any other implement.

The trowel, then, comes first; to be followed by a hand-fork, a small border-fork, a rake, a hoe (a draw-hoe is best on the whole), and then a

good useful pair of pruning scissors. The last-
named will serve for cutting flowers, as well as
for the small occasional prunings which are all
that most ladies attempt. Lastly, if you like to
indicate now and then to some pet creeper the
way in which it ought to go, provide yourself with
a stout hammer as heavy as you can use without
inconvenience, for there are few tasks more hopeless than to drive cast-iron nails into mortar with
the absurd dolls'-house kind of hammer which
" the Trade" considers suitable for a lady's use.
In spite of the popularity of the trowel, it will
perhaps be well to say a few words as to its legiti-
mate uses. One *has* known a trowel taken to put
slack on the fires of the conservatories, etc., to
knock nails in with, to spread mortar for any
home-made repairs, or to stir up water in which
soot, or some other plant-stimulant, was dissolving.
A tool thus misused will, of course, come to an
untimely end, however well it was made to start
with. Such treatment ought not to befall the
trowel dedicated to the mistress of the establish-
ment. She should keep her tools for her own use,
and rub them, or cause them to be rubbed, clean
and dry after each time when they have done
service.

It is chiefly for potting that the trowel is wanted. The compost suitable to the nature of the plants to be potted will have been prepared beforehand by the gardener, with the correct proportions of loam, sand, peat, or leaf-mould, as the case may require. The drainage having been deposited at the bottom of the flower-pot in sufficient quantity, and moss above that (if necessary), the soil is added with due care. First comes a layer of more or less depth, according to the size of the pot, then the root is placed, the fibres spread out, the plant supported with the left hand, and the remaining space filled in with the trowel, and the soil pressed firmly round by the closed hand. The trowel is indispensable in planting on rockeries, where the " pockets " or interstices which hold the soil are shallow, and no larger tool is of any use. In bedding out bulbs or small plants it is also exceedingly convenient, and for the planting out of seedlings it is, in some instances, better to use a trowel than a stick, as the round hole made by a planting stick of any size is not always completely filled up with soil; and a novice in gardening sometimes loses a whole batch of promising seedlings through this simple fact, that the top only of the hole is filled up by the pressing of the soil round the stem, while below the tender young roots are left hanging in a vacuum, and depending there unnourished, meet their death, which is followed by the decease of the stem and leaves.

The hand-fork is preferable to the trowel for removing plants from one bed to another. It has a short handle like that of a trowel, and three or four short flat prongs. In lifting a small plant with its ball of roots there is much less danger of injuring the fibres than when a trowel is employed, for the sharp edge of the latter cuts them all off at the point where the operator decides to " lift " ; while the blunt and open prongs of the fork merely thrust the roots aside, and if the lifting is done slowly and gently, the deep running fibres will be drawn out of the soil instead of being severed. And this is no unimportant matter, for torn fibres mean damage, more or less permanent, to the whole plant. To a geranium, a pansy, or any shallow rooting plant it causes a temporary check only, but that may just involve the blossoming coming two or three weeks late. A rose, or any other deep-rooting plant, will recover much more slowly, perhaps never entirely, from such an injury.

Another purpose for which a hand-fork is most

suitable is the loosening of hard, caked soil in a crowded flower-bed. When weather is hot and dry, and bedding-out plants are in full summer luxuriance, hoeing among the tender foliage is impossible, even to the most skilful hand; but with a hand-fork lightly used, the soil can be loosened, while leaves and flowers are preserved from injury with the left hand.

This involves stooping, of course; but in these days of athletic training, when women take part in so many out-door sports, there are surely few who do not possess a muscular back, which is, after all, a better thing than that " cast-iron " one, " with a hinge in it," for which the American gardener sighed in vain. A delicate woman who really cannot stoop must confine her gardening operations to the conservatory, or the potting-bench. But a strong one will find that after a very few days the stiffness and fatigue caused at first by stooping over the beds will wear off, and until it does, the best remedy is " a hair of the dog that bit her."

A border-fork will be required only by an energetic worker who is not ashamed to dig. It is made like the large fork used by the gardener, but is narrower and lighter. It is the best tool to use in turning over borders that have rose bushes, flowering shrubs, or clumps of herbaceous plants occurring at intervals, or where there are permanent bulbs. Here and there a bulb may be " spitted " and brought up on the prongs of the fork, but it can be replaced, and is not likely to be much injured ; whereas a spade or trowel would have cut it through, when " not all the king's horses nor all the king's men could put the two portions together again."

A rake is necessary for levelling the surface of a bed after digging or forking, breaking up small lumps and rendering the soil fine, and for removing pebbles, dead leaves, or weeds after hoeing. For use in the flower garden, a rather short-headed rake is best, with teeth set tolerably close together. There is scarcely any need to describe this familiar implement further, though perhaps the amateur may like to be informed that the price is regulated by the number of the teeth. The iron head is sold by itself, but is fitted to the handle before being sent home. A lady should choose a light, but fairly long, handle, so that, if possible, she can rake the borders without stepping on them.

It may not be inappropriate to remark here that detached beds, round, oval, etc., should be worked from the centre outwards, and long borders from the back to the front; otherwise the part supposed to be finished first gets trampled on while the other is being forked or raked, and the work undone. So in planting, a circular or square bed must be planted from the centre outwards; a border from the back, beginning with the taller plants, and ending with the edging in front.

A hoe is always a rather dangerous tool in unskilled hands. One feels so sure it has been thrust only just far enough to uproot that perky little groundsel ; and, lo ! there is a nice young aster beheaded, or a branch of a fuchsia fallen prone. Push-hoes are generally recommended for ladies' use, chiefly because it needs less exertion to work them ; but on the whole draw-hoes are safer. With the first tool, as the name indicates, the action is from instead of toward the operator, and is really less under control than a movement towards one's self. A fork is said to have been described as a " split spoon " ; it would not be a bad description of a draw-hoe to say it is like an unsplit rake.

In hot, dry weather the lover of her garden will find incessant uses for her hoe. It should then be called into play early in the morning, and a few hours' exposure of their roots to the sun will cause

the weeds to shrivel and die. Weeds grow larger
and faster in wet seasons, it is true, but they do
not mature and run to seed with the same rapidity
that they do in drought. If left lying about in
the rain they keep fresh, and soon contrive to take
root again ; so when hoeing is done in damp
weather the weeds must be quickly collected by
means of the rake.

In the use of the scissors there are many points
to be observed, even in the apparently simple
matter of cutting flowers for use in the house.
Many women who do not care for working in the
garden, yet like to cut flowers themselves for the
filling of their bowls and vases. Often they may
be heard complaining of the gardener's disobliging
nature, for the poor man cannot altogether conceal
his disgust at the ruthless style in which his plants
are hacked and hewn, and the way in which many
future blossoms are sometimes sacrificed to that
single one on which the mistress has set her heart.
The cutting of roses will be referred to later on.
Of other woody plants it may be observed that the
cut should be a " clean " one, not jagged, made
slopingly, and just above a leaf-bud. There will
then be no untidy bare end left ; a new spray
will quickly grow out of that bud, possibly with

fresh flowers upon it. In cutting geraniums, marguerites, dahlias, and other blossoms of soft-stemmed plants, the flower-stem should be cut where it joins the main stem ; unless, indeed, it is so short that some of the main stem must needs be cut too, in which case it should be taken off at a joint or above a bud. If the stem is too long for the vase to be filled, it can be easily snipped to the right length afterwards ; by taking it off at the joint both unsightliness and loss of vigour to the plant are avoided. When a plant has many blossoms, those to be gathered should be selected with some regard to the general appearance of the plant or bed; if a dozen or twenty are cut from one spot, then not only will there be an ugly blank, but injury may result to the branches or sprays so severely mutilated.

There are, of course, many other small articles which may be needed by the gentlewoman in her garden and glass-houses—a weeding-basket, nails, tags, a syringe, watering-cans, sponges for washing, small brushes for removing blight, larger ones for keeping clean the stages, floors, etc. Cleanliness is all-important in the successful management of plants, and the preservation of it is, or should be, a congenial task to feminine fingers.

" The sapless branch
Must fly before the knife ; the withered leaf
Must be detached, and where it strews the floor
Swept with a woman's neatness. . . .
Strength may wield the ponderous spade
May turn the clod, and wheel the compost home ;
But elegance, chief grace the garden shows,
And most attractive, is the fair result
Of Thought, the creature of a polished mind."

The woman who is to make order lovely in the
garden should herself be neat—neat, but most
certainly not gaudy. If she is really in earnest,
and devotes some time each day to her work, it
will be worth while to have not only a gardening
apron and gloves, but a particular gown for the
purpose. A dress that has once been smart and
fashionable is certain to have something about it
which is unsuitable for working in—even if it is of
the style of the year before last. The serviceable
gardening frock may be plain to prudishness, or
as daringly picturesque as the wearer chooses, but
it must have certain qualifications—the skirt must
not be skimpy, nor must it have its anterior fulness
tied back with tapes, which will either crack and
tear the stuff when the wearer stoops, or will
catch in her boot-heel (however low) when she
rises. The bodice should be made without bones,

the sleeves be tight, and of medium length. The material should be woollen, so that it can be brushed when dirty, and of a kind that does not cockle when damp. Reseda beige makes a capital summer gardening dress, and nothing beats serge for winter—grey, brown, or navy blue for choice. A crimson flannel blouse would give colour if worn with the latter, and the former may be brightened by a bright tinted apron, for the apron must be washable. A coarse holland or linen is best, and the self-coloured linens now sold for embroidery make charming aprons, boldly worked in cross-stitch. A " butcher's blue " is delightful with the reseda beige, or a deep red, or even orange for those who can stand it. A small square pocket on the right side (rather high up) is nice for one's pocket-handkerchief ; across the front, just at a convenient level to slip one's hand in, a fairly deep pocket, divided into two compartments, is most useful. In these can be put any small articles wanted handy for one's work—labels, string, wire, nails, tacks, raffia grass, pins for pegging down plants, and so on.

A pair of sleeves like the apron, to pull on over the dress sleeves, are also convenient ; they should reach to the elbow.

As to gloves, probably few of our readers would care to garden without them ; so it is necessary to warn them that if they really "mean business" gloves will be the most frequently recurring expense in all their paraphernalia. The gardening glove has yet to be invented, the finger-tips of which will survive three or four days of real effective work. No matter how well the glove is made, nor how strong the leather, the things are useless directly one's finger-ends peep through ; worse than useless, indeed, for nothing is more uncomfortable, in a small way, than to have pellets of soil work in at these holes, and trickle down into the palm of one's hand.

So for once the cheapest article is the most economical. Thick doeskin, or other gloves used for country walks, may, when soiled, finish their short existence in the garden ; and if between times it is necessary to buy, the most inexpensive glove procurable will be the wisest purchase.

CHAPTER III.

THERE are few people in these days who are content to work by rule of thumb. Cer tainly no educated minds care to do so. The cry for Higher Education has now been supplemented by that for Technical training; even cookery is taught on scientific principles.

Sometimes it is urged against gardening that it is a pursuit for stupid people only, affording little occupation to the mind, and still less occasion for exercising the intellect. It is perfectly true that many gardeners of the old-fashioned type, without any claim to book-learning, were successful in growing well-flavoured vegetables and a profusion of hardy fruits and flowers. They were, it is certain, possessed of a natural shrewdness, which enabled them to draw inferences from the experi-

ence of years. In most cases, also, these men had
at their command unlimited supplies of stable or
farmyard manure, and so they were able to produce
good results ; just as an old-fashioned country-
house cook, innocent of French dishes, could prepare most appetising bills of fare with the fresh
eggs, cream, fat poultry, and home-cured hams, all
ready to her hand.

We have endless horticultural difficulties to con-
tend with at the present, many of them of our
own creating. We like to have strawberries in
February, seakale in December, peaches in May,
peas in September. Every quarter of the globe is
ransacked for floral treasures ; and all the loveliest,
rarest, most delicate plants must be made to gratify
our eyes and senses in the variable climate of this
little island, where, as the Americans say, we have
not weather, but only samples.

If we will not let Nature take her way, then we
can but combat her by Science. Engineers build
for us artificial heated constructions, in which we
can force the strawberries and peaches. Botany
has gone hand in hand with horticulture till their
joint efforts have supplied us with peas that bear
in September ; by regulated temperature, artful
composts, condensed manures, etc., we manage to

grow the most unlikely plants under the most unnatural conditions.

The lady amateur gardener may or may not interest herself in all these matters ; but if she is active of mind as well as of body, the work will become tenfold more interesting when she has some insight into its scientific aspects.

Say she studies botanical construction alone (and it will last her some time), she will then know why this plant should be moved in spring, that in autumn ; if by some sudden accident—*e.g.*, as inhalation of bad gases—a plant loses its leaves, she knows which functions are interfered with, and what danger awaits the plant. It is more interesting to watch the growth of seeds if one understands something of the vegetable cell, or if the study of root-structure makes it plain how moisture or plant-food is taken up, and in different degrees at different seasons ; why deep planting is now condemned ; what is the object of some rules as to the potting of plants, and so on.

It is not every plant-lover who has the gift of what Mrs. Ewing, in her charming " Letters from a Little Garden," called " lucky fingers." She says there are certain people who may put in roots, cuttings, or seeds anyhow, and at any season, and everything grows without let or hindrance. But many of us may take pains to learn the right way, and then not always succeed the first time, or yet the second. To the uninitiated, probably, nothing seems easier than planting. You make a hole, put the roots in, and push the soil round them,— that's all.

Well, roots so treated *may* grow, but that is doubtful ; and they will most certainly be much retarded in their growth, and will blossom weeks later than if they are carefully planted.

You make the hole—true enough ; but its size must be regulated by that of the intended occupant, and if this has a root well furnished with fibre—as all healthy roots ought to have—then the circumference of the hole must exceed that of the fibres when spread out. The roots must be extended in all directions, not dropped in as a mass, and squeezed together in filling up. After they are spread out, the network must be filled in by letting soil trickle through the fingers down into the interstices, and then the remaining space filled in with hand, trowel, or spade, and finally the surface pressed firm about the top of the roots. In some plants, all the leaves and flower-stalks spring from one centre, called the " crown " ; this should never be covered with soil, but allowed to project just above the surface ; otherwise, when the soil is moist it will cause the crown to rot away, ending in the loss of foliage and blossoms. Of this habit are the primrose and all its cousins ; cyclamens and others springing from corms or tubers ; the Japanese anemones, funkias, and many more.

Geraniums, fuchsias, and all plants which branch out from main stems, should not have much of this stem placed underground, else it is liable to rot, and again away goes the head of the plant. Shallow planting is an article of the creed of modern gardeners ; but, of course, the term is comparative. What is shallow for a heliotrope would not suit a bush-rose, and a standard rose or fruit-tree must go deeper in proportion than bushes ; while plants that make large bushy heads must be lower in the ground than those of slenderer habit, or their own weight, during some storm or gale, may drag them up by the roots. The object of this somewhat new rule of shallow planting is twofold. First, because the roots, being near the surface, form their fibres there. The fibres or root-lets are the parts that derive food and moisture from the soil. When close to the surface they can

benefit by every slight shower in summer, and
by manures (or plant-food) applied at the top.
Secondly, they are easier of removal, the roots
being in a compact mass. Deep planting caused
the formation of one or more long main roots (or
underground-stems, as the case may be), which
wandered off to the right or left, as attracted by
some suitable soil for feeding in. When it is
necessary to move such a root as this, the
gardener must dig down and down, following the
course as well as he can, often cutting into the
main roots as they take some unexpected turn, and
then finally at the end hangs the fibre, which
probably is half torn away by the difficult process
of getting at it. When advertisements appear
offering trees, shrubs, or roses at prices that seem
attractively cheap, the reason is sometimes given
that the owner is thinning out his plantations, or
something of that kind ; and such plants, if
bought, will generally prove to have just these
long straggling main roots, with a wretched little
bunch of fibre dangling here and there, and it is
a remote chance that they will ever be able to
establish themselves again when replanted. To
the novice, it may seem a recommendation that the
shrub or tree in question had never previously

3

been moved ; in reality, it is nothing of the sort. When about a year old, a tree or bush should be moved, even if only from one part to another of the same bed or garden, the object being to prevent the roots straying, and to keep them in a compact mass.

There are, of course, some exceptions to this rule of shallow planting. In a dry, hot, sandy soil, for instance, almost everything may be planted deeper than is wise in heavy, moisture-retaining land ; especially is this true of spring bulbs or corms, and roots planted in early spring. Those light soils are generally early ones, and the shoots would start up too quickly through the loose earth, and perhaps be injured by frosts, or forced into premature and therefore weakly bloom. The gladiolus, by the way, is rather fond of a fair depth of rich soil to root into. Lilies have the same predilection. Among popular shrubs the rhododendron is one that roots deeply; and indeed it is hopeless to grow these beautiful plants in a shallow soil.

The usual mistakes of the amateur in potting plants are insufficient drainage and want of firmness. It is better to use a number of very small bits of broken crocks, rather than two or three large pieces, and they should cover the bottom of the pot for a depth of a quarter of an inch to two or three inches, according to the size of the pot. It is pretty safe to assume that our readers will not have to mix their own composts ; this will be done for them by the gardener, also a proper supply of clean pots provided. Pots and drainage must both be clean, else the health of the plants will suffer. Pots should not only be clean, but also dry. The soil, on the other hand, must be neither dry nor wet, but just moist. Wet soil not only clings unpleasantly to the hands (or gloves) of the operator, but it will not be worked in sufficiently among the roots of the plant. Dry soil cannot be made firm at the time of potting, and seldom or ever becomes so afterwards. The reason why there is more potting, or repotting, to be done in spring, is that as the plants are then starting into new growth, fresh soil and more or less additional space for the new rootlets is advisable.

" More or less space," because while soft-wooded plants take no harm by being moved into a pot several sizes larger than that they previously occupied, it will probably kill a hard-wooded plant —a heath, azalea, or camellia, to be so treated.

Even with the soft-wooded plant, its flowering season will probably be retarded by being " shifted " on so far at once ; because a plant growing in a pot generally makes no growth above ground until the rootlets have grown through the new soil, and reached, or nearly reached, the side of the pot.

If the plant to be repotted is an old one, the soil must not be entirely broken away from its roots ; it must be removed in a "ball," as gardeners say— that is, with all the earth which is (so to speak) in the clutches of the roots, left intact. And in planting a ball of roots, care must be taken that the new soil is pressed home close up to the ball, or else the roots will never strike into it, and the plant will be rather worse off than before. For the same reason, a ball of roots should not be placed quite dry in the centre of new moist soil.

A well-potted plant will bear the test of being turned upside down and given a fairly sharp little tap without leaving its new quarters.

The remarks on shallow planting hold good to a great extent with regard to seed-sowing. While large seeds must be sown comparatively deep, very small ones do best when merely sprinkled on the top of a fine bed or pan of soil. The seed, for instance, of tuberous begonias should be thinly scattered at the top of a pan of light soil containing plenty of sand to keep it open. As the minute seeds would be swept away by the gentlest of waterings with a fine rose, the necessary moistening of the soil is accomplished by placing the pan inside a larger one filled with water until its soil is saturated. Large hard seeds must be soaked in luke-warm water before they are sown, more especially if they are not new. Cucumber and marrow seeds germinate much more quickly if this is done, and with many seeds brought from abroad it is quite necessary. It must be remembered that there are some seeds which lose the power of germinating after a very short time ; others will retain this power for years. It is a wise precaution, in using garden seeds left over from the year before, to try a few of each sort beforehand, and thus prove which are in condition for use. This saves much waste of time and trouble in preparing and sowing seed-beds which may come to nothing.

Hardy annuals, such as sweet peas, may be sown out of doors in November, but rather deeper in the soil than spring-sown seeds would be. But it is a disadvantage to sow half-hardy seeds a long time before they are wanted ; the hardy ones lie under the soil through the winter, and germinate at the natural time in spring, but the others

requiring artificial protection are forced into speedy
growth, and then, as it is not the season for planting them out, there they must stay, in house,
frame, or pit, crammed together in their pans or
boxes, and kept, maybe for months, at a standstill.
Many of these seedlings will die ; the remainder be
drawn and weak.

Beds for sowing out of doors must be forked, and
then raked till the soil is as fine as possible; and
for raising seeds in boxes, etc., very light mixtures
of soil with a large proportion of sand is desirable.
The seeds of wild plants do not fall into holes or
cracks, but most likely sprout under the light
covering of dead leaves, and such *débris* as may be
blown over them, or washed by rain. The chest-
nut or acorn in the woods gets no other preliminary
covering than this, however solid a mass of heavy
soil the tree may eventually embrace with its roots.

Those who like saving their own seeds should
take some little trouble in selecting the finest
flowers, and marking them, so that they shall not
be gathered by mistake. If there is a particularly
fine plant from which seed is to be saved, some of
the inferior blossom should be cut away, and only
the best left to go to seed. Seeds must be collected in dry weather, and stored in a cool dry place.

Propagation by the division of roots is practised with regard to the numerous herbaceous plants now so popular in the flower gardens. October or March are the two months most .suitable for dividing and removing this class of plants ; in the neighbourhood of large towns it can be safely carried out only in the spring.

Growing from cuttings and from layers are the only other methods of propagation very likely to be resorted to by our gentlewoman in her gardening experiments.

A common mistake in taking cuttings is to choose shoots that are either too young and soft, or too old and dry. To take the familiar instance of bedding geraniums. Every one knows how soft, light in colour, and *juicy-looking* are the young shoots, and that towards the close of summer the older ones get quite hard, brownish, and woody in appearance. Neither of these will strike satisfactorily. It is the middle-aged shoots, so to speak, that are wanted ; they are not too full of moisture, like the first, nor have they lost their vigour, like the second. In short, as the gardener would say, it is the half-ripened wood that makes the best cuttings.

Cuttings—or " slips," as old-fashioned people

called them—must be severed from the parent stem by a clean sharp cut, slanting downwards, and just below a bud, where buds on the stem exist. In setting cuttings, care must be taken that the soil is pressed round the stems, right to the end. Not infrequently, whole batches of healthy-looking slips will damp off; and if the soil is turned out it will appear that when the first little thread-like root came forth, it found itself in a cavity, and so, having no support for its weakness, it died, and presently the stem followed suit.

Layering is most familiar as practised on carnations. A cut, as if for taking a slip, is half made, then the shoot bent downwards, and laid in soil placed in a little mound conveniently close to the parent plant; the layer is held down by a little peg, the spray standing up as if growing; roots are put forth by the half-severed stem, and when established the young plant is finally cut away from the old one. Tree carnations may be layered thus: a sort of palisade of small sticks is put round the edge of the pot, outside which a wall of stout brown paper is tied. The addition then made to the pot is filled with light soil, and the layers fastened down into it all round.

Tuberous plants, such as dahlias, may be increased by cutting the roots into pieces, each of which must have one or more " eyes," and then it will make a perfect plant.

Bulbs and corms are increased by off-shoots ; but as they will have a chapter to themselves, directions how to treat the baby-bulbs will be given there.

CHAPTER IV.

THE FLOWER GARDEN.

THE perfect garden is that in which there is to be found something sweet and pleasant every day of the year. "The first knots, or buddings of the spring," are, perhaps, the dearest of all, because of their freshness. As Leigh Hunt says: "One of the greatest pleasures of so light and airy a thing as the vernal season arises from the consciousness that the world is young again; that the spring has come round; that we shall not all cease, and be no world." In summer the plentifulness of beauty is apt to spoil us. We become *blasé* and indifferent, and rather welcome autumn with its subtler *fin-de-siècle* effects, wherein beauty and decay so strangely mingle. Sweetest of all, however, are the winter flowers that bloom so bravely in the snow, and give colour and life to the torpid and dreaming earth.

One of the surest signs that a lady takes personal interest in her garden is that the flower-circle of the year is complete there. The ordinary gardener, if left to himself, will devote all his energies to making a great display in summer ; though he may be induced to give some of the space in his borders, formerly occupied by his endless " bedders-out," to herbaceous plants. Of course if the owners never, under any circumstances, visit a country house except in late summer, and if economy is an object, it would be mere waste of time and expense to grow flowers for making the garden gay at other seasons. But such cases are now very rare. Flying visits are often paid to country homes at all seasons, or they are lent to friends, or to young couples in their honeymoons.

Stiffness of design, with hard and violent contrasts of colour, is apt to mar the laying out of a garden where no cultured taste has any influence. A mistress who may be physically unable to attempt the planting of her own beds, may wisely direct the manner in which this will be done. Above all, let her plead the use, in summer especially, of sweet-smelling flowers. The most gorgeous and effective of our modern favourites lack the charm of fragrance.   In spring,

"Violet banks, where sweet dreams brood,"

are the pleasantest bits of the garden.   Frost
and snow cannot dismay them.   Lift the umbrella-
like leaves, and you will find the tiny flowers
beneath, white and stiff with rime, perhaps, but
as sweet scented as ever.   The hyacinth is one
of the few florist's flowers which has a perfume.
But its smell is not an *English* one—it is too
rich and heavy.   Rather does it breathe remi-
niscences of its Eastern home, and we associate it
more readily with the song of the bulbul, and
"the fragrant bowers of Amherabad," than with
an English garden.   It is one of the harem of
night-flowers which Moore speaks of as wooed by
the Spirit of Fragrance, for its scent is strongest
at eleven o'clock at night.   There is a prevalent
idea that this stiff and stately flower owes its
origin to its pretty cousin the graceful bluebell
of the English woods ; but this is not so, nor do
botanists now even place them in the same genus.

Among spring shrubs none are more grateful
than the lilac and the honeysuckle.   They are
the first to put forth green shoots; and as early
as Christmas, in sheltered corners, the honey-

suckle will unfold its leaves. The leaves of
the lilac are later, but it and the poplar are
green and summer-like when most of the other
trees are bare and brown. There is a delicacy
about the perfume of the lilac, a grace about
its drooping plumes of flowers, which help also
to make it a favourite. Like the hyacinth, it
(with many other good things) comes from the
East. The Persian Ambassador at the Austrian
Court of Ferdinand I. introduced it along with
the tulip; and now in mid-Europe the lilac runs
wild in the hedgerows, and here in our English
gardens it grows as if upon its native soil. It is
a very charming cosmopolitan. Its botanical name
(*Syringa*) has led to occasional confusion with
the syringa (vulgarly and unworthily nicknamed
the "mock orange"); but this mistake has not
even the excuse of relationship.

As fragrant and more effective in masses is
the hawthorn, "the sweet-smelling May," a veritable native of the land. It has always been
associated with Royalty. For many a year it had
its peasant queens in every village; it was the
insignia of the Tudors, who adopted it after
finding the crown of defeated Richard in a hawthorn bush; it was the favourite of the French-

bred Queen of Scots, and till lately Queen
Mary's thorn bloomed yearly by the still waters
of Duddingstone, which sleep hard by the grey
Salisbury Crags, not far from the Forth estuary.
Most recently of all, the May has been associated
with a Princess in whom all Englishwomen must
ever take a tender and reverent interest, inasmuch
as on the eve of her bridal her betrothed was
taken from her and hurried "into the silent land"
alone. This was to have been the year of the
May *par excellence*, but now few of us can look
upon the pretty flower unmoved.

But all these are to the undiscriminating male
gardener as nought, compared with his scentless
and brilliant begonias, gladioli, geraniums, abutilons, cannas, marguerites, and dahlias. Of all
these, the marguerites are perhaps the only
plants of purely British breed. The geranium
has numerous English relations, and may be
considered a naturalised foreigner. The pelar-
gonium is, however, a decided colonial, and hails
from the Cape of Good Hope. The begonias are
tropical invaders, and have somewhat of an alien
look on a smooth English lawn, whilst the abutilon
comes from Brazil. The gladioli come from
the Cape of Good Hope, and the dahlias from

Mexico, although the latter plant is, curiously enough, named after a Swedish botanist. Spanish travellers brought it to Madrid about the time of the French Revolution, but it was not until after Waterloo that it was fairly at home in our borders. Against this terrible invasion of foreigners we would protest. A certain proportion of them may be admitted as ambassadors, and their brilliant uniforms will give state and distinction to the floral court, but they should not be allowed to oust wholly our sweet-scented, old-fashioned English natives.

What says Bacon in his essay " Of Gardens " ? " And because the breath of flowers is far sweeter in the air (where it comes and goes like the warbling of music) than in the hand, therefore nothing is more fit for that delight, than to know what are the flowers and plants that do best perfume the air."

It is worthy of note that this association of the perfume of flowers with music occurs in Shakespeare also :

> "That strain again—it had a dying fall.
> O, it came o'er my ear like the sweet south,
> That breathes upon a bank of violets,
> Stealing, and giving odour."

" Roses damask and red " (continues Bacon) " are fast flowers of their smells ; so that you may walk by a whole row of them, and find nothing of their sweetness ; yea, though it be in a morning's dew."

Tea-roses, with their seductive beauty and far-carrying odour, were then unknown ; but, alas, the accusation brought against Elizabethan roses holds good concerning all the most showy hybrid-perpetuals of the Victorian age.

Yet for all our neglect of fragrance, we can improve upon the Baconian list of flowers that scent the air. Bacon names first "the violet, especially the white double violet, that comes twice a year—about the middle of April and about Bartholomewtide." This, by the way, is a lost variety, for we have none now that blossom in August. " Next to that," he continues, " is the musk-rose. . . . Then the flower of the vines. . . . Then sweetbriar. . . . Then pinks and gilliflowers, especially the matted pink and clove gilliflower. Then the flower of the lime-tree. Then the honey-suckles, so they be somewhat far off."

To these we may add hyacinths and narcissi—

> " Fairest among them all,
> Who gaze on their eyes in the stream's recess,
> Till they die of their own dear loveliness "—

syringa, white jasmine, with its prim, sweet,
Quaker-like flowers, and stiffly decorous foliage,
Clematis flammula, *Magnolia grandiflora*, heliotrope, with its nun-like habit and subtle perfume,
Japanese and other half-hardy lilies, some of
the tobacco-plants, and many others, not forgetting
that universal·favourite, mignonette. One always
feels as if mignonette were really an English
flower, it has fitted so naturally into our affections ; and it is difficult to believe that its first
home was in Egypt, beside old Nile. Perhaps
the perfume-loving Cleopatra plucked and used
it in the days when she enchanted brave Antony
by the river of Cydnus.

In autumn the supply of scented flowers falls
rather short. Our tea-roses linger on with faint
sweetness as long as the weather keeps fine.
Russian violets bloom again in October, and even
in depths of winter. Sometimes a sudden perfume may surprise one in passing an old-fashioned,
insignificant plant, the *Tussilago fragrans*, known
as the winter heliotrope. This tussilago is a
South European cousin of the brave little colts-
foot, that sends its golden cyclops-eye forth into
the cold spring, and before even its own green
leaves will venture to accompany it.

4

The gardener may be allowed to do his best (or his worst) with devices in scarlet, purple, yellow, and blue, and carpet patterns in dwarf foliage plants in the beds upon the lawn, or rather the edges or ends of the lawn ; for it is pure vandalism to cut up the green expanse by spotting it over with circles, triangles, and octopus-like patterns.

In a place apart, open but not exposed, are the rose-beds, if there be no rose-garden ; for the "royal rose" blooms best alone. And in some sheltered dell or warm nook sub-tropical plants may rear their stately heads. But the special charge of the lady of the house should be the long borders of herbaceous plants, so beautiful in themselves, so interesting from association ; and specially delightsome because she is able to cut and come again just as fancy impels her.

Here, in spring, will be found the "rathe prim-rose," the daffodils that come "before the swallow dares," the graceful and delicate hepaticas (both pink and blue), Leopard's-bane (*Doronicum Harpur Crewe* is the best), globe-flowers, the "luckie gowan" of the superstitious Highlanders, and the earliest "dew-cups of the frail anemone," especially the white *A. sylvestris* and the brilliant *A. fulgens* ; fritillary or snake's-head, some creamy-white, others with the strange chequered markings that have caused the country people to call them guinea-hen or turkey-hen ; jonquils, slenderest and most girlish of the narcissi; *Dielytra spectabilis*, with its pretty pink, heart-shaped flowers pendent from the arching stem. It is known in some parts of the country as " Lift-up-your-head-and-I'll-kiss-you " (a very insinuating title), and in others, less appropriately, as " Bleeding hearts." The present fashion of potting up this flower for the sake of getting it into blossom a few weeks before time is a very foolish one. It is not an effective plant under glass; it goes off very soon,.and is not half so robust as when grown out of doors. Then, large pots must be used, which might more profitably be occupied by some tenderer, rarer specimen.

Nothing exhibits a more wonderful range of colour than wallflowers, and nothing more effectually scents the air. The wallflower is now considered to be the gillyflower of most poets, but not of all ; for many undoubtedly mean by this pretty old-fashioned name the carnation. Tusser, however, talks of

" gillyflowers all,
That grow on the wall,"

and many and varied are the quaint and pleasant
fancies that circle about the name. Cowley says
the gillyflower was

"Jove's flower, when Jove was but a child,"

and Clerk Sanders declares—

"Their beds are made in the heavens high,
Down at the foot of our good Lord's knee,
Weel set about wi' Gillyflowers."

Good Sir Walter Scott was antiquarian enough
to share this old-time enthusiasm; for he said
he was sure that he would find the gillyflowers
growing in heaven.

The fact that both the wild wallflower and the
Dianthus caryophyllus—the original clove pink—
flower in July on old walls, and have "speckled
flowers," may account for the considerable confusion that has undoubtedly prevailed as to which
is the real gillyflower. Bacon and Drayton at
least meant the carnation.

Herrick has a pretty legend concerning the
wallflower, which runs thus—

"Why this flower is now called so,
List, sweet maids, and ye shall know.
Understand, this firstling was
Once a brisk and bonny lass,
Kept as close as Danäe was;

Who a sprightly springall loved.
And to have it fully proved,
Up she got upon a wall,
Tempting down to slide withal :
But the silken twist untied,
So she fell, and, bruised, she died.
Love, in pity of the deed,
And her loving, luckless speed,
Turned her to this plant we call
Now "the flower of the Wall."

This brings us to the spot where spring' and summer meet—a season at which the herbaceous borders should be full of beauty. Clumps of the dear old white and pheasant-eyed pinks should scent the air; double potentillas will be opening their gay balls, and the iris—the *fleur-de-lis*, as some think, of France—is there in all its nationalities, Spanish, English, and German. "Ancient heralds," says one writer, "tell us that the Franks of old had a custom at the proclamation of their king to elevate him upon a shield or target, and place in his hand a reed or flag in blossom instead of a sceptre." Along with these come the charming little *Spiræa filipendula* and the blue veronica. The latter perpetuates the pretty legend about St. Veronica ;— how when Christ bore His Cross up the hill, she alone was ready with a cloth for Him to wipe away His perspiration with. Look close into the flower, and in the white-marked centre you will see an indistinct resemblance of a human face. The Germans call the flower " Man's Love," for you have only to blow it ever so gently—away it flies.

> " Sigh no more, ladies, sigh no more,
> Men were deceivers ever ;
> One foot in sea, and one on shore ;
> To one thing constant never."

Now, too, come larkspur (*Delphinium*) and monkshood. The larkspur is in every shade of blue. Its German name—knightspur—is a better one, for its fantastic blossoms closely resemble this. As for monkshood, it is a gloomy-looking plant, as all poisonous plants should be. If you pull off the hood you will find that the nectaries inside bear a certain resemblance to a car drawn by doves.

When the foxgloves and bell-flowers come they sound the *carillon* of summer, and the time of abundant bloom has arrived.

> " The poppy fair that bleeds
> Its red heart down its leaves."

comes later, and with it its sisters in every dainty combination of colour possible. The " Oriental " and the " Shirley " will be found most desirable.

The latter is an annual, but we must admit some
annuals into our mixed borders—blue cornflowers,
annual chrysanthemums, and sweet peas, if no
others. Nor must such perennials as double pyre-
thrums and columbines be left out—the latter if
for no other reason than the joy of seeing its
dainty green rosettes of leaves unfold themselves
in spring. Its flowers are said to resemble a nest
of doves—hence its name. The white scented
mallow (*Malva moschata-alba*) is very attractive,
and so is the plume poppy (*Bocconia cordata*), on
account of its decorative capacities. There is something very charming about the evanescent beauty
of the evening primrose. *Œnothera speciosa* has
large white scented flowers, and does not receive
half the favour as a border-plant that it deserves.

Late summer will turn the penstemons into
full beauty ; but the kings of autumn are the
hollyhocks. They stand, like generals in their
brilliant uniforms, high above their fellows.
Japanese anemones (admirable for artistic uses),
cannas, and dahlias come then also, and the
summer-flowering chrysanthemums herald the
passage into autumn.

The signals of autumn make us sigh, not that
autumn is anything but lovely,—only because it is

the last of the three gladsome seasons. Yet
often in September the herbaceous borders will be
at their bravest. All these, with the sunflowers
and autumn star-worts, or Michaelmas daisies, make
a grand show. *Pyrethrum uliginosum* (for which
as yet no English name has been found), and many
others, will last on far into October, or even
November, weather permitting.

But at last the day comes when nothing remains
but to cut down and clear away the shrivelled stems
and blackened foliage, and to make all as tidy as
may be ; adding, perhaps, a thin coat of leaf-mould
and rotted manure on the surface of the bed. Care
must be taken not to disturb the roots ; for when
*once a herbaceous bed is well established, the less
interference with roots the better.*

In December there should be some Christmas
roses (or black hellebores), to save the bed from
utter desolation. If this be indeed the hellebore
of the Ancients, the grand remedy for madness
and hypochondria, it is a very pretty one, and
an excellent cure for December hypochondria, at
least ; for who could come upon the charming
flower, with its snowy petals, without feeling that
life was brighter and healthier than before ?

CHAPTER V.

THE ROCK-GARDEN.

"OF all forms of cultivating flowers, rock-gardening," it has been said, "is the most fascinating," and there will be few dissentients among those who have tried this form of cultivation.

The rockery is achieved by the tradesman at his suburban residence, or is a perpetual joy to the retired sea-captain, who adorns it with huge shells and battered figure-heads, and erects a flagstaff in the middle. It needs but the outlay of a pound or two on stones and soil, and a few days of hard but willing labour, and the thing is done.

Far otherwise is it with the making of a rock-garden. This entails the expenditure of much thought, money, and time. It cannot be built in a day, still less be immediately peopled with rare and fragile beauties ; but once established, it needs no great exertions to keep it in order, and the constant small attentions it requires may be undertaken by a lady who does not care for much stooping, or for the more arduous garden tasks, which she will probably distinguish by the name of " grubbing."

A rock-garden may be made by utilising a natural valley, or by scooping out a miniature one, and varying the sides with undulating hillocks and hollows. Or a third plan is to raise small Alpine ranges on the flat, and this requires much more careful handling than the others, or it will be so stiff and obviously artificial as to repel rather than attract.

The rock-garden at Kew may be taken as a model of the charms of irregularity of arrangement. It has been formed by the scooping-out process, the advantages of which it illustrates very happily. But then Kew has to be instructive, even before it may be beautiful ; and so the plants—which, for the sake of the student, must be named—are grown, for convenience' sake, in patches; a plan that need not be followed in a private garden. The site was occupied as lately as 1882 by a flat lawn ; and the luxuriant, well-established look which has now reigned for some years should be encouraging to those who are contemplating the laying out of a rock-garden in this style.

Water, in some form, should be introduced, if it does not exist; for the marsh plants, which are so beautiful and interesting, cannot be grown otherwise.

Memories of two rock-gardens pass before the mind's eye as we write ; one in the grounds of a pretty old house in the Midlands, the other not ten miles from Charing Cross. Both these are made by throwing up mounds of soil and stones on a flat surface ; and, strange to say, the suburban owner has succeeded in producing a charming and artistic effect, while the old country garden is spoilt by the unnatural rampart of rocks which borders an artificial stream, cutting in two the wide lawn that-used-to-be *by an almost straight line*.

The other is simply an unusually large villa-garden, with the ordinary fence dividing it from the road, and the familiar red brick villa in the midst. But good taste, ingenuity, and loving care have contrived to devise a rock-garden, which, in the variety of its tiny inhabitants and its skilful arrangement, causes the real to approach very nearly to the ideal.

In such a garden let us wander for a little while, and note what are the most pleasing and interesting of its features.

As we go along the winding walks, shaded
and diversified by evergreen oaks, clumps of
flushed or snowy azaleas, and plentifully blooming
rhododendrons, we come here and there on bits of
elevated ground, where art effectively simulates
nature. Low shrubs grow on the top, and hardy
rock plants nestle in the pockets, looking as if they
had never known another home. The next bend of
the path brings us to a miniature tableland, over
which creepers crawl and tumble with the blithe-
some intrepidity of a baby. At the foot of the
tableland may be a shallow pond, upon whose
shores water-crowfoots and sundews may grow at
will. No rock-garden is complete without the
sundew—most curious and quaint of native plants.
Swinburne has sung the praises of the

> " Little marsh-plant, yellow-green,
> And pricked at lip with tender red " ;

and says that

> " The deep scent of the heather burns."

About it ; its natural barbaric apparatus for catch-
ing flies, and its strangely fleshly look, combine to
make it seem more like some savage alien than a
home-bred citizen of our "native heath." In a deep
corner, near at hand, there may be tall ferns, sturdy

irises, tritomas, with drooping sword-like leaves and
aspiring heads, and the wonderfully symbolic reed.
Who can look at it without remembering Mrs.
Browning's vivid version of the ancient legend ?

> " He cut it short, did the great god Pan,
> (How tall it stood in the river !),
> Then drew out the pith like the heart of a man,
> Steadily from the outside ring,
> And notched the poor, dry, empty thing
> In holes as he sat by the river."

Further on we shall look for a rocky mound,
where sedums may grow in all their variety. Some
of the names of these prim, well-behaved little
plants are very expressive, such as " Tricquema-
dame " and " Prickmadam." They are the types of
thrifty housekeepers, for they will store up the
moisture of one shower, so that it suffices to keep
them green and fresh for months in the driest
seasons and most exposed situations. Many
another choice Alpine and American rock-plant
may be chosen to keep them company. Spiky
yuccas should crown the next ridge ; whilst hawkweeds, saxifrages, pinks, and daphnes variegate the
interstices. There is a meaning in the name of the
hawkweeds, for the keen-eyed birds are reported to
" cleere their sight by conveying the juice hereof

into their eyes." The saxifrages are all lovable little plants, though in a garden they do not charm by their unexpectedness as on the barren Highland hills.

The chief interest in the daphne lies in the pretty story of the nymph of that name whom Apollo pursued. It is probable that it was the true laurel which the Greek imagined to be the transfigured nymph; but it is the spurge-laurel that bears her name. The late Mr. Russell Lowell, in his "Fable for Critics," made very witty use of the story. But the swift-footed daughter of Lado has ever been a favourite of the poets. Carew (who speaks of her as the bay-tree, by the way) pictures her freed from her woody prison, and hanging on the Apollo "like his Delphic lyre."

> " Full of her god she sings inspiréd lays,
>   Sweet odes of love, such as deserve the bays,
>   Which she herself was."

Comus threatens his unhappy victim with the fate of Daphne, and Byron sings—

> "Yield me one leaf of Daphne's deathless plant."

But this is wandering far away from our rock-garden, and we will leave others to settle whether the laurel, the bay, or the spurge-laurel be the

home of the nimble maid. Round many flowers
there hang pretty memories like these ; and it
is pleasant to write and think of the ancestry of
our favourites.

A stony bed should be provided for the lovely
but difficult gentian. Give it but a sufficiently
ascetic home, and it will grow thick, and strong,
and beautiful. Felwort and baldmoney are its old
names, and Chaucer speaks of it as baldmoyne.
Gentius, King of Illyria, was the first to discover
its medicinal uses; hence its name. Emerson calls
it the "blue-eyed pet of blue-eyed love," and Jean
Ingelow talks of "God's gentian bells."

This is an ideal rock-garden ; and if it is worth
while to have one at all, it is surely worth while
to have it as perfect as possible. Now for a few
directions as to management ; then we must pass
on to the plants.

Care should be taken to build none of the
rock-mountains too steep, otherwise the moisture
runs off, and in dry weather no plants can live—
unless it be a few hardy stonecrops, which are not
of the rare quality we desire to cultivate. The
actual nature of the soil thrown up to make the
mounds is not of great importance, because only
enough of it should be used to afford the stones

a solid bed. The " pockets " in which the plants are to go should each be filled with soil adapted to the wants of the plant at the time when it is inducted in its new habitation. And of this soil, whatever it is, a fresh dressing should be added every spring. There should always be plentiful supplies of water obtainable for use in hot weather, when, by employing the hose, everything can be kept in verdure and beauty. Such washings, as well as those of the rain, will always carry away a certain amount of the soil in the pockets, and hence the need for yearly replenishings. For most bulbs, and such plants as grow in woods or pastures, a mixture of loam and leaf mould will suit ; distinctly mountain plants, heaths, rock-roses, daphnes, etc., will want peat, the gentians stiff stony soil ; the lovely little hardy *Cyclamen ibericum* and its variations require a sprinkling of old crumbled mortar every spring, to be followed by a good soak of water.

The mistress of a rock-garden will have herself to blame if she cannot here find some floral gem to admire, at whatever season she walks through it. True, it may be only some tiny Alpine, whose delicate beauties need a microscope to reveal them fully ; but there *should* be flowers always, sometimes showy, sometimes not. Indeed, those who love show, and who grow flowers for cutting alone, had better not attempt rock-gardens ; they will lead but to disappointment.

Foremost, in the bitter month that "freezes the pot by the fire," the *Cyclamen Atkinsi* will be found. The cyclamen is a member of the primrose family, although its petals, turned back like rabbits' ears, give it a rather dissimilar appearance. It has some curious habits, and as the seed ripens the flower-stalk curls round it, and so defends it from the winter cold ; nor do the seed-vessels spring open and distribute themselves till summer. There are many ancient superstitions concerning the cyclamen, and the corms, dried and powdered, were believed to act as a love-philtre.

Soon after this come the squill, the "glory of the snow," and its cousin, the *Scilla Siberica*. Squills, like gentians, are well-known medicinal plants, and Gerarde has one quaint use for them : "The inner part of squills, boiled in oils and turpentine, is with great profit applied to the chaps or chilblains of the feete or heeles." Hepaticas (single and double, pink, white, and blue) are early in the year, and *Adonis vernalis* (the pretty lady cousin of the homely buttercup) is

charming, not only for its flower, but for its name.
Ancient herb-women called it Rose-a-rubie, and the
French *Goutte de Sang*. Some say, however, that
it is really the anemone that enshrines the spirit
of the youth whom Cytherea loved, and Shakespeare
seems of this opinion :—

> " By this the boy, that by her side lay killed,
> Was melted like a vapour from her sight ;
> And on his blood, that on the ground lay spilled,
> A purple flower sprang up, checkered with white."

But the colour of the adonis makes it seem to us
the rightful owner of the personality. The

> " Fair frail anemones, which starlike shake,"

may grow here also, however. *A. fulgens*, *A.
nemorosa bractea*, and *rubra* are the best for the
rock-garden. *Aubrietias*, some of the choicer
daffodils, primulas, polyanthi, the pretty American
cowslip (*Dodecatheon*), and auriculas (dusty millers,
the country people call them) are all suitable
plants. Alpine pinks, lovely and fragrant, will
bloom in early summer, and last on for weeks.

Whilst speaking of odorous plants, we must not
forget to include the creeping evergreen daphne
(*Cneorum* or "garland flower"). Asphodels, the

most floral of rushes, are charming for early
summer, and there is the strangely-fashioned
columbine,

> " which maids forsaken
> Ever in their garlands wear."

You may have their elaborately constructed flowers,
whose singularity gives them distinction, in scarlet,
yellow, and azure (*Aquilegia californica, chry-
santha*, and *glandulosa*). The *Funkias* are handsome, and grow well in warm, dry spots, and
the *Heucheras* are valuable for their foliage
(like many a modest dame, who, save for her
marvellous confections, would be nought). One
species, however (*H. sanguinea*), adds the charm
of character to that of clothes, inasmuch as it
has a beautiful flower (crimson) as well as fine
foliage.

In passing, we must record a protest against the
brutal tastelessness of botanists, who persist in
dubbing every other crimson flower " sanguinea."
It has unpleasing associations with the butcher
and vivisector ; and though we do at times dissect the dainty things, and let their juices flow,
let us dissever them, as far as possible, from all
unpleasant thoughts.

The graceful and slender *Lychnis* (in its dwarf Alpine varieties), the stiff and formal saxifrage (in its rarer manifestations), and the somnolent poppies, may all combine to maintain a succession throughout the year. The Iceland poppies, *Meconopsis*, *Aculeata*, and *Wallachi*, are all good. The *Wallachi* (known as the blue Indian poppy) is a beautiful plant, which grows tall and strong in a cool, moist place with peaty soil. But whether from Iceland or India in reality, the true dream-home of the poppies is Lotus-land. We cannot see their fragile, fluttering leaves, swaying lightly with every breeze, without wishing we too might drift as lightsomely through life, and

> "Dream and dream like yonder amber light,
> Which will not leave the myrrh-bush on the height."

The poppy recalls many memories, but the pleasantest is that of Endymion and

> "The magic bed
> Of sacred dittany and poppies red,"

where he had such strange, sweet visions. All his story, indeed, breathes the faint, uncanny smell of the poppies, that hang

> "Dew-dabbled on their stalks."

But where all are charming it is difficult to choose, and, even as it is, we have left many beauties unmentioned. There are Alpine phloxes (there are, indeed, a bravery and delicacy about all Alpine plants that give them a special charm) and bell-flowers, that tinkle joyously and noise-lessly all day long; the *trillias*, the day-lilies (that bloom early and plentifully), *spiræas*, *sedums*, *sempervivums*, and the tiny Alpine honeysuckle, about two and a half inches high, but sweet in proportion to its size as the English woodbine,

"Of velvet leaves and single blooms divine."

Nor must we forget to mention (though we have not space to describe their beauties and peculiari-ties) the American hardy *Cyprepediums*, or ground orchids.

Autumn is by no means a barren time. There are the "hot pokers," contrasting in September so beautifully with the soft greys and lavenders of the starworts. Of these a judicious selection may be made to last from July to November; *e.g.*, *Aster Alpinus* (July), *A. Polyphyllus* (August), *A. Cordifolius* and *Dumosus* (September), *Eri-coides* (October), *Grandifloras* (November). The graceful Alpine *Anemone Pulsillata* will flower in autumn—with those who succeed in making it flower at all. The European cyclamens flower in early autumn; and at the same time the *Plumbago Larpentæ* shows its deep blue petals, the *Polygonum affine* its rosy spikes.

In November, *Gaultheria procumbens*, a dwarf evergreen shrub, will be bright with scarlet berries; also *Iris fœtidissima*; and the golden wintercress should don its brightest colour.

In the next month, the Christmas roses (*Helleborus niger*) will creep out, and furnish masses of snowy blossoms, if protected from wet and wind by a bell-glass. Of the hellebores a selection may be made, and last on from December to March. The *Cyclamen Ibericum*, a native of the Caucasus, in a sheltered position, will often flower in December.

Needless to say, we have been obliged to pass over many decorative foliage plants, as well as old-world favourites, which will occur to every one, such as the mystic St. John's wort, the curative veronicas, and sunflowers, emblems of constancy. Note the tiny black spots all over the perforated St. John's wort. These were made by the needle of the "auld Mahon." It seems a

strange occupation for " auld nickie ben," and
there is a prevalent idea that he has not many
leisure moments ; yet it is only another proof of
the fact that the diversions of the eminent are, as
a rule, very singular and unexpected.

CHAPTER VI.

THE WILD GARDEN AND WATER-PLANTS.

THE wild garden should deserve its name from two distinct points of view; first, because in it many of our native plants are grown; secondly, because everything there, whether British or foreign, must grow at its own sweet will, and in a position as like to its natural place of abode as circumstances will permit.

It is useless to lay down rules for the formation of a wild garden, because no one deliberately sets to work to create one; rather does the wild garden develop gradually out of some well-directed effort to beautify a waste spot. It may be a straggling shrubbery, bordering a drive, or a strip of rough ground, with grass and trees, between the garden proper and a wood or planta-tion; or again, it may be a steep bank, grass-grown, and dotted with bushes, which is too steep or uneven to be kept in the fine, close-shaven trimness of the neighbouring lawn.

"Nothing will grow there under the trees," is the common complaint; till at last it strikes some one that as primroses thrive in the woods, they might condescend to flower on the bank, or in the shrubbery, and the children delightedly go off to forage for a supply, the roots are dibbled in, and lo! next spring the bare spot is lovely with "vestal primroses." There are no flowers more loved of poets and of men, and so our pleasure in them arises not only from the exquisite delicacy of the flowers, but from the many pretty and dainty fancies that cluster round them. They are of the flowers that Perdita longed to give her lover.

Milton talks of

"The rathe primrose, that, forsaken, dies."

And even Wordsworth, although as a rule a gallant upholder of the gladsomeness of the woodlands, says that

"The patient primrose sits
Like a beggar in the cold."

The joyous Herrick chides it for its melancholy—

"Speak, whimpering younglings, and make known
The reason why
Ye droop and weep.
Is it for want of sleep,
Or childish lullaby?

> Or that ye have not seen as yet
> The violet?
> Or brought a kiss
> From that sweet heart to this? "

But surely the sight of the hardy yet delicate flowers that carpet the woods, when still the trees are bare and brown, is anything but a grievous one. Wordsworth speaks of them another time as

> " Glittering at evening like a starry sky."

Carew has the prettiest conceit of all in connection with it :—

> " Ask me why I send to you
> This primrose all be-pearl'd with dew ;
> I straight will whisper in your ears,
> The sweets of love are wash'd with tears.
>
> * * * * *
>
> Ask me why the stalk is weak
> And bending, yet it doth not break ;
> I must tell you these discover
> What doubts and fears are in a lover."

The success of the primroses will encourage you next year to try the

> " Cold snowdrops, which the shrinking new-born year
> Sends, like the dove, from out the storm-tost ark."

Ruskin doesn't like the snowdrop. He calls it a " tiresome flower," and says you " nearly break

its poor little head off, and take all sorts of
trouble with it, and then half of it is not worth
seeing." Certainly they are not flowers for general
effect. " Pendent flakes of vegetating snow "
some one has called them ; yet they are scarce
plentiful enough to have the effect of snow. It
is indeed their timorousness, their loneliness
when they step so bravely out of the hard brown
earth, the first-comers of the year, that endears
them to us. Wordsworth has an exquisite picture of them,—

> " Who fancied what a pretty sight
> This rock would be if edged around
> With living snowdrops? Circle bright!
> How glorious to this orchard ground!
> Who loved the little rock, and set
> Upon its head this coronet?
>
> *   *   *   *   *
>
> It is the spirit of Paradise
> That prompts such work, a spirit strong,
> That gives to all the self-same bent,
> Where life is wise and innocent."

The crocus follows naturally in the wake of
the snowdrop, and the daffodils tread close on
the heels of the crocus. The Rossettis had a
weakness for the crocus. Its brilliant flame-yellow
appealed to their sense of colour. Dante calls

it a "withering flame," and Christina tells how the

> "Crocus fires are kindling one by one."

But of all flowers none, save the rose, has been so beloved by the poets as the daffodils. They filled their cups with tears when Lycod died, and they were another item in the choice tribute Perdita longed to yield her lover.

> "Daffodils,
> With the green world they live in,"

were among Keats's "shapes of beauty." The older poets loved it, and the bank by Cynthia's well was "with daffodillies dight."

Herrick has immortalised them in the loveliest of his lyrics; and Wordsworth had never a finer vision than that of the

> "Host of golden daffodils,
> Beside the lake, beneath the trees,
> Fluttering and dancing in the breeze."

> "Fair, frail anemones which, star-like, shake
> And twinkle by each sunny bank or glade,"

will also prosper under the trees, and ferns will unroll their young leaves right joyously here. Smaller than the wood anemone, yet like it in the veined delicacy of its petal, is the wood-sorrel. It is called in Savoy the "bread of God,"

because it is so plentiful.   It is the primula, or
first flower of spring of Norway; and in Italy and
Spain it still goes by its old name of " Alleluia."
" Alleluia-flower " the English called it, and possibly
do so still in country districts ; while " Cuckoo's
bread and cheese " is the children's name for it
in some of the western shires.   Gerarde called it
" cuckoos' meat," and says,—" Either the cuckoo
feedeth thereon, or by reason when it springeth
forth and flowereth, the Alleluia is sung in the
churches."   Its root creeps " like beaded coral,"
and its leaves shrink when roughly handled, or
when a storm is brewing.

Nor amongst early spring flowers should the
lesser celandine be wholly forgotten.

Thus the wild garden may start life in the
spring ; but the spring is only the beginning of
things, and there is no reason why the seasons
should not be as fitly and fully represented here
as elsewhere.

When the first spring flowers are beginning to
flag, regiments of wild hyacinths should be ready to
take their place, and their natural companions are
the stitchworts and the earlier and more fragile
Umbelliferæ—the slender earth nut, the graceful
cow parsley, etc.   The cow parsley is a particularly

good thing to have beneath the trees, for its delicately and elaborately cut leaves begin to spread
out in January, when there is comparatively little
as fresh and green as they. The slim, erect habit
of the bluebell improves by contrast with the wild,
irregular graces of these others. The fritillary is
a less common native ; but Matthew Arnold has
localised it—

> "I know what white, what purple fritillaries
>   The grassy harvest of the river-fields,
>   Above by Ensham, down by Sandford, yields,
>   And what sedged brooks are Thames's tributaries."

Another uncommon yet truly indigenous plant is
Solomon's seal. It owes its name to certain black
spots at the base of its petals, which resemble a
Hebrew symbol, supposed to have been used by
Solomon as his sign-manual.

One often hears it said that the walnut-trees
"poison the ground," and nothing will grow under
them ; but this is a mistake, for cowslips, daffodils,
stars of Bethlehem, and *Tulipa sylvestris* (an exquisite little yellow tulip that still grows wild in
some parts of England), all seem utterly unconscious
of this baneful influence. The last-named plant is,
moreover, very fragrant, and therefore to be prized.
The perfume of the first spring flowers is very

delicate. The primrose has its odour when you
hold it up to your face, but few are sufficiently
odorous to scent the air. Loveliest of scented
plants are, of course, the lilies of the valley—

> "Shading like detected light
> Their little green-tipt lamps of white."

The *mai-blume* is full of sentimental and practical
associations.

> " Valley lilies, whiter still
> Than Leda's love,"

Keats calls them, showing truer taste than Shelley
in his familiar description—

> " The naïad-like lily of the vale,
> Whom youth makes so pure and passion so pale,
> That the light of its tremulous bells is seen
> Through their pavilions of tender green."

What, indeed, has passion to do with this most
vestal of flowers—the " virgin's tears," as it has
been called ?   But there is an inexplicable tastelessness about Shelley at times.

Clare has a pretty picture of the

> " Stooping Lilies of the Valley,
> That love with shades and dews to dally,
> And bending droop on slender threads,
> With broad hood-leaves above their heads,
> Like white-robed maids in summer hours,
> Beneath umbrellas shunning showers."

Another wild woodland flower, valuable for its odour, is the woodruff, but, like "the actions of the just," it smells sweetest after it is dead (or dried) and hung up. The Germans put it into their charming drink *bowle*; and it is a principal ingredient in the exquisite bouquet of many foreign liqueurs. In its fresh condition, though not so odoriferous, it is very pleasing with its glossy, star-like whorl of leaves and its tiny, twinkling flowers.

Pretty in its way, but terribly distinctive in smell, is "ramsons," or wild garlic. It is another plant that will grow healthily beneath walnut-trees. Its leaves somewhat resemble those of the lily of the valley, and the flower is pretty enough, but the smell is a sad disenchantment. It belongs to the genus *Allium* (the onion tribe), a branch of the great lily family. The onion family, although it has ceased being proud, and associates now too freely with the plebeian and inscrutable sausage, was once a very important one. It is difficult to think of the onion as " a daughter of the gods "; yet it has had divine honours paid it in Egypt and other countries.

This wild spot may also be made a very pleasant garden of memories if we bring flowers from every different part of the country we visit. Yellow

poppies from Wales, heath from Cornwall, lilies of
the valley from Cambridge, " virgin's bower, trailing
airily," from Kent, cranesbill from the Severn, *Iris
fœtidissima* from the Devonshire lanes, the blue-
bells of Scotland from their native heath;—and even
heather from the Highland hills may in some
suitable spot " flame a purple deep as dawn."

Honeysuckle should climb over low bushes,
hedges, fences, or some other support of natural
appearance. The clematis, before alluded to as
" traveller's joy " and " virgin's bower," should not
be omitted. Flowers, fruit, and foliage are all alike
decorative.

Where the grass absolutely refuses to grow under
trees, it often happens that periwinkles can be got
to make a close green carpet, especially the smaller
species. " Freshe Pervinke, bright of hew," it is
called in the " Romaunt of the Rose " ; and it is
àpropos of this flower that Wordsworth made his
charming confession of faith—

> " And 'tis my faith that every flower
> Enjoys the air it breathes."

Lysimachia nummularia (moneywort, or " creeping
Jenny ") is another useful plant for this purpose.

On a bank in a shrubbery some brambles should
be grown for the sake of their splendid autumn

colouring. There is nothing as "plenty as black-berries," nor as beautiful. In January you will see the new leaves shooting forth, while still the old ones droop soberly from the prickly stems. Very sweet and pretty, too, are

> " The bramble blooms
> That fill the long fields with their faint perfumes,
> When the May-wind flits finely through sun showers,
> Breathing low to himself in his dim meadow bowers."

And as for the berries, the " Ethiops sweet," as Emerson calls them, they are lovely to look on, and lovelier still to eat. Here, too, may meadow-saffron, the delicate lilac autumn crocus, send up its pretty flowers—

> " Like lilac-flame its colour glows,
> Tender, and yet so clearly bright."

It was, moreover, an ancient medicine; and at Saffron Walden, in Cambridgeshire, a special culture of it was inaugurated in the reign of Edward III. It is used now mostly as a dye for cooking purposes, the orange-coloured stigmas being the only part really employed. It is a plant of such peculiar habits as to be worth studying. The seed-vessels form underground, and do not rise until spring, when they appear above ground, ripen in the sun, and disperse themselves.

Sometimes the wild garden grows up round the edges of a pool ; or there may be a pool in the grounds which can be converted into a water-garden of itself. If so, there is great scope for the exercise of taste and ingenuity; for not only can water lilies and other beautiful plants be utilised, but grasses, reeds, iris, and so many graceful objects can be grouped upon the banks, making them picturesque and luxuriant.

Yet to those who desire to grow and study that class of plants which nurserymen dub " Aquatics," the possession of a pond is not a *sine quâ non*. Water plants have been grown with entire success in a series of wide tubs filled with water. We remember calling at a house in the south of England where there was nothing whatsoever that was commonplace, from the owner downwards, throughout every corner of his house and garden. As regards the latter, every bed was a storehouse of rareties and curiosities ; and as we drove out at the back (to take another road homewards), we marked a row of tubs (about two and a half feet high and three and a half feet wide) ranged along the wall of the house, each having two or three " aquatics " in a high condition of salubrity.

The following list, it must be understood, is one

of perfectly hardy plants, suitable for growing out of doors in this climate, not for ponds and fountains under glass.

The "floating water-lilies, broad and bright," naturally hold the first place. They are said to be the handsomest flowers in all the British flora ; and they are, in addition, very queens in song and story.

They encircled the island home of the lily of Astolat, and shared their name with that lonely and hapless maiden. There is an exquisite picture in *Comus* :—

> " Sabrina fair,
> Listen where thou art sitting,
> Under the glassy, cool, translucent wave,
> In twisted braids of lilies knitting
> The loose train of thy amber-dropping hair."

Mrs. Hemans yearned for water-lilies to " cool her fevered brow " (it is noticeable that our poets are less " fevered " than of yore ; indeed, some of them are so astonishingly callous that a trifle of the earlier " frenzy " might not be amiss).

Water-lilies are the delight of the poets and the despair of the scientists. Botanists classify, dissect, and learnedly discourse concerning them; yet the delusive beauties still keep their secrets hid in

their golden hearts.   No one can explain how and
why they open only in sunshine, and always sink
below the surface of the water at sundown.   Of
these habits Tennyson makes excellent use,—

> " Now folds the lily all her sweetness up,
>   And slips into the bosom of the lake ;
> So fold thyself, my dearest, thou, and slip
>   Into my bosom, and be lost in me."

Hardier, commoner, and less striking than the
first, is the yellow water-lily (*Nymphæa lutea*),
which has a strange spirituous odour, which gains
for it in some country places the prosaic name of
" brandy-bottle."

The author of *Olrig Grange* writes enthusias-
tically concerning

> " Its egg-cup, yellow and full,
> Just outside the fringe of sedge."

There is also a small yellow lily (*N. Pumila*),
which grows in some Highland lakes, and even in
such civilised waters as the Mugdock and Bardowie
lochs that lie just outside Glasgow.   *N. Rosa* is a
perfectly hardy water-lily, a variety of *N. alba*, with
rose-coloured petals.   A newly introduced kind
is *N. Marliacea Chromatella*, large and canary-
coloured, with spotted leaves.   Water villarsia (*V.*

nymphoides) is found in England growing wild,
but rarely. It is very like a yellow water-lily,
but the flowers are flatter, and fringed. A white
variety is sometimes procurable; it is strong and
pretty.

"Frogbit" is the strange name of a rather
uncommon water-plant with creeping stems, heart-
shaped leaves, and very delicate white flowers,
which is found in England and grows best in
stagnant water. "Water Soldier" lives in ditches
etc., in the Eastern counties, and when in flower
floats altogether on the surface of the water, and
sinks to the muddy bottom afterwards ; it has stiff
prickly leaves, like those of an aloe, while the
flower is white. *Hottonia palustris*, water-violet,
is well known; it has pink and yellow flowers,
small, but arranged in whorls round a stalk which
rises some inches out of the water. A variety is
obtainable with pale-blue flowers. The handsome
pink blossom of the flowering rush makes it
a desirable inmate of the pool, or water-garden ;
this plant grows in canals and slow rivers, and
is not uncommon. It is curious how many of
these water plants are pink. The water-plantain
is another, not so handsome as some of its comrades,
but with the meritorious habit of keeping long in

bloom. Again, there is the arrowhead (*Sagittaria sagittæfolia*), both leaves and flowers of which are very pretty, while it is marvellously hardy, and is said to linger on in the Thames, quite in its mid-course through London.

CHAPTER VII.

HARDY CREEPERS AND CLIMBERS.

EVERY one is more or less fond of that class of plants generally designated " creepers " ; and everybody's ideal of a country house or cottage is one whose walls shall be partially hidden by a network of branches, luxuriating here and there into tangles and trails of greenery and flowers. There should be " a rose looking in at the window," eglantine round the door, jasmine rampant on its front, and ivy climbing up the back, to embrace the chimney-stacks.

The odd part of it is how very small a list of " creepers" most people are content with. To judge by appearances, any person ignorant of the subject might well be excused for thinking that nothing could be induced to " climb " in this country, except roses, jasmine, ivy, Virginian creeper, and the *Jackmanii* clematis.

As a matter of fact, the plants which can be used

for covering walls are so many and varied that
there is ample room for choice, if only people would
take a little more trouble.   There may not be any
tangible excuse for a woman who adopts an ugly
or unbecoming style of dress, just because all her
intimates are wearing it ; but her conduct, even if
not excusable, is at least comprehensible to her
fellow-women.   They know either that it is a
necessity of her nature to copy others, or that she
lacks the independence and energy to withstand the
decrees of her milliner, who, for the time being,
clothes all her customers alike.

No excuse, however, presents itself for suffering
one's garden to be a mere slavish imitation of
some one else's or everybody else's.   In the choice
and judicious use of uncommon " creepers " there is
every opportunity for a lady to show her taste, and
impress some degree of individuality upon her walls
and boundaries.

Before naming any of the plants suitable for
these purposes, it is well to point out that,
correctly speaking, " creepers " will not cover
walls, neither will " twiners " ; climbing plants
alone are fitted by nature to perform that part.
The plants which are in common parlance *lumped*
together as " creepers " really fall by their natural

habit into three divisions,—climbers, twiners, and creepers.

A true climber attaches itself to a wall or other support by means either of tendrils, stems, or roots. Such are the vine, clematis, passion-flower, *Ampelopsis*, etc. The twiner grows spirally, round and round some other stem or suitable object, as the hop does, or the convolvulus. It is interesting to note that each particular twining plant always grows from the right, twisting round to the left, or *vice versâ*. The creeper, properly so-called, literally creeps or grovels along, generally rooting into the ground here and there along its course. You may plant it on a height, and it will grow downwards, as if trying to reach the earth ; but it is quite contrary to its nature to grow upwards. In technical language, creepers are procumbent, or trailing ; as are periwinkles and tradescantias, called after Tradescant, the celebrated gardener of James I. Some of the *Crassulas, Euonymus radicans,* and *Muehlenbeckias,* belong to this class.

It is possible that the poets are accountable for some of this confusion between twiners, creepers, and climbers. There is one reckless description of the " sky-blue periwinkle " climbing " e'en to the cottage eaves," for instance ; and it is in points like these that we are enabled to differentiate between the true lover and student of nature, and the vague and inaccurate sentimentalist. You do not find Wordsworth writing at random. His description of how the periwinkle " trailed its wreaths " shows his acquaintance with it.

Of modern climbers, there are none more attractive and varied than the clematis family, all of them colonial and foreign cousins of the graceful wayside plant already referred to as Traveller's Joy and Virgin's Bower. The clematis is a plant to which deep rooting is somewhat of a necessity. It likes good soil also, and a fair amount of manure, particularly in liquid form. It is quite useless to buy some of the beautiful varieties which are to be had nowadays, and then to plant them in a dry, worn-out border or in a gravel walk, as is sometimes the case. Those sent out by our nurserymen have generally English names; being hybrid varieties, from original types. As there are about thirty-two original species known, it will be imagined how many are the varieties, although it is only a little over thirty years since the first hybrid was produced by a Scotch gentleman. There are eight classes into which horticulturists divide these thirty-two kinds, but many of the distinguishing features are not such as call for mention here. Only these broad facts should be borne in mind by any one going in for clematis culture—viz., that those plants which are marked in the trade catalogues as *Montana type*, *Patens type*, or *Florida type*, bear their flowers on the ripe wood, and therefore must not be cut back on the approach of winter. Those that blossom on the newly formed shoot of the same season belong to one or other of these four—the *Graveolens*, *Lanuginosa*, *Viticella*, or *Jackmanii* types. Some good kinds belonging to the first class are Lady Londesborough, silver-grey in tint, blossoming early; Miss Bateman, white with brownish centre; Lord Derby, lavender ; Duchess of Edinburgh, white, double, and sweet-scented ; and Lucie Lemoine, white with pale yellow centre. These last two need a warm sheltered position. To the hardier kinds which blossom on the young shoots, the standing favourite, *Jackmanii*, belongs. Others to be recommended are Blue gem, *Lanuginosa candida*, Othello, Thomas Moore, *Flammula rubella*, and Star of India.

The clematis has climbed so effectually into our affections that it is hard to tear ourselves away from it. A whole volume has been devoted to it by two admirers of the other sex ; and we might well give a chapter to its praises, only that would be scarcely fair to the many other beauties whose charms we desire to enumerate.

The *Eccremocarpus* is a very graceful climber, which scarcely receives the attention it deserves, as both its orange-scarlet flowers and light green foliage are very decorative. Admirable effects may be obtained by allowing light-foliaged deciduous plants, such as this and the wistaria, to intermix with the darker-hued evergreens. *Eccremocarpus* and *Cobea scandens*, the Passion-flower, and *Aristolochia sipho*, all require a southern or sheltered western aspect. The *Cobea scandens* comes from Mexico, and is a very rapid grower, with large bell-shaped flowers, green at first, and purple afterwards. The *Aristolochia sipho* is called also pipe-shrub, pipe-nine, or Dutchman's pipe. Its drooping flowers have a curiously crooked and inflated perianth, with reddish-brown veins, which is supposed to resemble a pipe. It is a native of the Alleghanies, and a cousin of the curious little native birthwort (*Aristolochia clematitis*). As for the Passion-flower, it is perhaps the most charming of all. The Jesuit missionaries in South America used it to teach the simple natives the mysteries of the Cross and Passion; the different parts showing the number of the Apostles, the rays

of glory, the nails and the hammer. It has been
used to emphasise a less sacred passion than this.
It climbed round the lion that " ramped " on Maud's
garden gate, and it wept a " splendid tear " when
Maud delayed to join her lover. In one of Jean
Ingelow's poems there is a pretty picture of it.

> " The thatch was all bespread
>   With climbing Passion-flowers ;
> They were wet, and glistened with raindrops shed
>   That day in genial showers.
> ' Was never a sweeter nest,' we said,
>   Than this little nest of ours.' "

There are many rather hardy flowering shrubs
which flourish exceedingly when trained against a
wall, and can be made to serve the purpose of
climbers, although by nature they assume the bush
form.

Among these are the *Budleya globosa*, with its
gay little orange balls of blossom, and the Japanese
privet, which does not succeed everywhere out in
the open, but is very effective, and produces an
abundance of its white spikes of flowers against a
warm wall. The " superb magnolia " (*M. grandiflora*) is without rival among these shrubs. Its
delicious fragrance (the " musk of magnolia ") is
only one of its charms. It is a veritable Yankee,

and " faint was the air with the odorous breath of
magnolia blossoms " by the lakes of Atchafalaya,
when Evangeline passed by. Every one knows it ;
but some are not aware that there are other magnolias, hardier, if less magnificent, which, after the
fashion of the insignificant, will often flourish where
the *grandiflora* will not. One with rather dull
purple flowers is even accommodating enough to
grow against a north wall. The magnolia and the
Japanese privet are both evergreens.

The same monotony is shown in the choice of
evergreen as of deciduous creepers. *Pyrecanthus*
and *Cotoneaster* are repeated on so many walls that,
could one ever really dislike anything so harmless,
we should dislike these, and their perpetual red
berries. *Escalonia*, with its waxen bells, is scarcely
ever seen ; nor the *Ceanothus azurea*, with its small
close foliage and pretty blue flowers. The *Argara
microphylla* has fern-like leaves, and sweet-scented
though inconspicuous flowers. Of ivies the choice
is very large, and nothing is more continually satis-
factory. " Wanton Ivie " Spenser called it, but
it is the very type of friendship and constancy.
Garrya elliptica is also worth growing, if only for
the sake of its long handsome catkins. There
is a charm about all " tassel-hung " trees, a grace

about their pendant flowers, that we can scarcely define.

The twiners, such as the whole honeysuckle tribe —the trumpet, French, Dutch, evergreen, and variegated—are often grown on walls, but are better suited for trellis arches or stumps of trees. There is no sweeter plant anywhere than " ye woodbine hanging bonnielie." Beatrice ("dear Lady Disdain") hid her in a

> " Pleached bower
> Where honeysuckles, ripened by the sun,
> Forbid the sun to enter."

A sentimental and unpractical poet speaks of it as

> " Recompensing well
> The strength it borrows with the grace it lends " ;

but it is very injurious to trees, particularly young ones. Its graceful whorls of florets served to give hints to skilled Roman and Grecian architects, and, of late years, decorators have awakened anew to its value. It has a faithful lover—the white admiral butterfly—who, under pretence of smelling its flowers, feeds greedily on its leaves.

The tuberous tropæolum and perennial peas are also charming. Hood says—

> " The pea is but a wanton witch
> In too much haste to wed,
> And clasps her rings on every hand."

The best peas are *Lathyrus latifolium splendens, Albus*, and *Azureus*. *Meteor* is a very good tropæolum.

In conclusion, we must name a few valuable flowering plants, which, being only half-hardy in this climate, need to be planted out for the summer only, and to be protected, in a cool greenhouse, during the late autumn and winter ; or in very favourable aspects some of them will stand out if ashes, straw, or dried fern are placed above the roots. By the way, similar precaution must be taken in cold, damp places with regard to the tropolæums and everlasting peas.

Maurandya Barclayana is a pretty plant, which may be put out in May. There are varieties with white, pink, or purple flowers, rather like an etherealised foxglove, which go on all through the summer. *Lophospermum scandens* is a half-hardy annual, which may be grown easily from seed, sown during March in a hotbed. Plumbagos will do well against a warm wall in summer, also abutilons, large old plants of heliotrope and geraniums, and, of course, the hardier kinds of fuchsias. The lovely and varied ivy-leaved geraniums really belong to this class of creepers or procumbents; but, with the support of a trellis, they can be made to climb—or rather to appear to climb—if it is desired.

There is no better way of obeying Leigh Hunt's exhortation to do our " best for the earth, and all that is upon it, in order that it may be thought worth continuance," than by cultivating liberally this charming class of long-armed and embracing plants. They are the most irrepressible beautifiers we have. The straight and arid expanse of oak-fencing may, by their aid, bloom with verdure, and the dull, dry wood almost dreams that it lives again. Nothing so becomes a brick wall as a climbing green thing. Many an ugly but useful building or boundary may, by the aid of these plants, be trans-formed into a part of the beauty around. Yet should the walls of the house be not too liberally bedecked. When covered with greenery they look very pretty outside, but inside it may lead to peel-ing paper, or black, mouldy patches on the wall. The austere, utilitarian housewife loves not creepers; yet may compromises be effected, and walls inclined to be damp be painted and varnished rather than papered. The varnish is not so readily affected by them, and will, moreover, act as a barrier between the damp and the interior. Thus may peace be made between the house-lover and the garden-lover, beauty may reign both outside and in, and the " continuance" that Leigh Hunt looks for be the lot of both.

CHAPTER VIII.

BULBS AND THEIR CULTURE.

IF people's ideas are limited on the question of climbing plants, still less are they extended on the subject of bulbs.

" My soil is so light and sandy that nothing will grow," sighs some amateur gardener.

" Why not go in more for bulbs ? " we gently insinuate.

" Of course we have crocuses and snowdrops," is the reply ; " and I have a small bed of tulips and hyacinths. One can't have many, you know; they are so dear."

These four families are supposed to represent the whole bulb tribe ; and the two last must be used sparingly because of their cost, so there is nothing more to be said or done.

This chapter is written to prove that a good deal more may be done, and that without reckless expenditure. Perhaps it will be best to reverse the

order of the title, and give a few general directions
on bulb-culture before recounting the variety of
bulbs which an amateur may easily manage.

We may remark here, in passing, that what the
unbotanical person calls a bulb includes, as a rule,
not only bulbs, but tubers and corms. The bulb is
not a root, but a subterranean bud. It is, in fact,
the centre of life in the plant. The thick scales
grow round it in layers, and defend its vital parts
from cold, etc., just as our ribs and flesh protect
the more essential organs within. Here, safe in its
warm home, it sends its stalks up to feed on the air
and light, and its fibrous roots downwards to drink
in the rich nourishment of the earth. Both the
superb lily and the humble onion work existence on
this plan. In *Lilium bulbiferum* tiny baby bulbs are
produced in the axils of the leaves, and these may
be detached and cultivated into fresh vital centres.

The tuber, on the other hand, is a subterranean
branch, which is arrested in its growth, and becomes
thick instead of long—just as a hump-backed human
being, having his normal growth arrested, develops
an abnormal excrescence instead. Potatoes and
dahlias are familiar examples of tubers.

The *corm* is a development of the tuber, and may
be defined as a dilated stem, intervening between

the first buds and the roots. Buds in the form of young corms develop laterally from the parent corm. The crocus and gladiolus are corms.

The first remark we want to make about bulb-culture will, perhaps, astonish some of our readers— *i.e.*, that it is a mistake to plant bulbs, even choice bulbs, in beds by themselves. Experience teaches that bulbous plants growing under trees, or at the edge of borders filled at the back with flowering and other shrubs, will flourish, increase, and flower year after year with renewed vigour, provided they are occasionally thinned out if they become overcrowded. This vigour is not to be accounted for by the fact that the roots are undisturbed, for bulbs out in an otherwise unoccupied space will often go off without having been interfered with. No ; the reason would appear to be that it suits bulbs to pass through their period of rest in soil that is occupied by other roots which are growing and feeding all the time that their neighbours lie dormant. The activity of these other roots keeps the soil sweet and whole-some, whereas ground that lies idle while the bulbs are resting becomes too strongly impregnated with certain elements to be beneficial to their growth when they start forth again.

Bulbs will grow in stiff soil, or in that which is

light and porous ; but it is not always wise to leave them so long undisturbed in the first as in the latter, because in damp seasons they are apt to become mildewed or diseased. In this case the badly infected bulbs must be taken up, and thrown away ; but those which are only a little damaged may have the bad parts cut out. The bulbs must then be washed, and allowed to dry thoroughly— in fact, they must be kept for several weeks before replanting.

Another point to be observed about the management of bulbs, is that their leaves need cherishing in every way at all periods of their existence ; if the leaves suffer the root will suffer also, and the formation of off-sets be checked entirely. Whether your bulbs be hardy or tender, give every protection you can to the young leaves. This does not mean that they are to be unnecessarily coddled, but that if a very mild November and December have encouraged irises and plants of that kind to shoot up small green spears, and if afterwards sharp frosts come, straw, matting, fern, etc., should be spread just to keep those premature leaves from being cut off.

Most bulbs do best without manure, but to those which it suits, it should be given in the dry form of an artificial manure, or as a liquid when the flowers are developing and need support. Stable manure must not be allowed to touch a bulb or corm, as it will cause fungus, or some other form of disease.

The scope of this volume does not permit of giving information as to the management of rare or delicate bulbous plants. Nor is it necessary to give general hints on the planting of the commoner bulbs, for these are to be found in every nursery-man's list, from which the purchaser will make his selection. So, having pointed out several facts not widely known, it is time to name some of the more interesting kinds of bulbs, both for outdoor use, and for culture under glass.

Many of the hardy sorts have already been men-tioned, as denizens of the herbaceous border, rock, or wild garden. But there are still plenty to recom-mend. It will be news to most people that there are now forty-two recognised varieties of the snow-drop. This calculation does not include the "pink snowdrop," the discovery of which in a garden near Norwich caused some excitement a few years back; but which (alas for the delighted horticulturists!) proved to be manufactured by some frolicsome girls who had watered the bulbs with Judson's dye! There really are two yellow snowdrops; and these, with several of the rarer white and green ones, have been rediscovered of late years in old gardens, either Scotch or in the extreme north of England; while of the rarer daffodils Irish gardens have been the careful guardians during the long period when the tribe was condemned as " vulgar " in this country. Now, no one can have too many of them, and the desirable kinds are too many to enumerate. Some of the most beautiful are not to be depended on everywhere, or by everybody, so perhaps the following list may be of service. It was compiled after the Daffodil Conference held in the spring of 1890, at the Royal Horticultural Society's Gardens at Chiswick. By general vote of experts, these seem to be thought the best and strongest for garden purposes.

Single yellow	{ Emperor. { Glory of Leyden.
Single bicolor	{ Empress. { Grandis.
Single white	Madame de Graaff.
Double daffodil . . .	{ Telamonius. { Poeticus. { Phœnix (orange and sulphur).
Single (large)	{ Sir Watkin Wynn. { Gloria Mundi.
Polyanthus	Grande Monarque.

Of summer-flowering bulbs there is *Hyacinthus candicans*, not very beautiful as regards its individual blooms, but very effective in the mass, especially when planted where its creamy spikes grow through a mass of some bright-flowering bushes, such as fuchsias. There is a variety distinguished as *Hyacinthus princeps*, but it is a poor one, and should be avoided. Then *Ixias, Sparaxis*, and *Babianas* are interesting, easy to grow, and very decorative. All have slender spikes of prettily coloured flowers, most useful for table decoration. Here it may be noted that most of these plants belong to one of three great bulbous families—the *Iridaceæ, Amaryllidaceæ*, and the *Liliaceæ*. The iridaceæ have mostly corms, rhizomes, and fibrous roots, and include the crocus, iris, and gladiolus. The amaryllidaceæ are bulbous, and number in their ranks the snowdrop, daffodil, American aloe, amaryllis, etc. The lily family is very large and comprehensive, is both bulbous and tuberous, and includes the asparagus, lily of the valley, *Dracæna*, hyacinth, tulip, Solomon's seal, onion tribe, squill, star of Bethlehem, fritillary, and many others.

The *Ixia* is of the iris family. Its name, which is Greek, and signifies bird-lime, was given to it because of its clammy juice. It was mentioned long ago by Theophrastus. The varieties of it produced by hybridisation during the last few years are wonderful and beautiful. There is one which is distinctly peacock-green in colour, unlike anything else in the floral world.

The *Bahianas* also belong to the *Iridaceæ*, and come from the Cape, although there is one species a native of Socotra. The Dutch gave them their name because the baboons ate them. This was feasting on a lily with a vengeance.

Gladioli are too well known to need recommendation, but we must put in a plea in favour of their being planted in clumps at the edge of shrubberies, or in mixed borders among tall plants.

The dear old St. John's or midsummer lily is without a rival in early summer; but how many beauteous forms have been introduced which flower in August and September—*Lilium auratum*, with its spicy odour—the golden-rayed lily of Japan ; the martagon lily, with its graceful habit and Turk's cap bloom ; the gorgeous Canadian lilies in flame-like yellow and red ; and *L. lancifolium*, called also *elegans* in recognition of its dainty grace.

In Shakespeare's days, the white and Chalcedonian lilies were most familiar. Both came from the East ; the white lily was the type of the Virgin, and the

scarlet Chalcedonian was generally thought to be the
" lily of the field." Shakespeare makes as liberal a
use of the lily as of the rose. When Guiderius enters
bearing Imogen insensible, Arviragus exclaims,—

> " O sweetest, fairest lily !
> My brother wears thee not the one half so well
> As when thou grew'st thyself."

Poor disregarded Katharine, Henry VIII.'s first
wife, says,—

> " Like the lily
> That once was mistress of the field, and flourished,
> I'll hang my head and perish."

And a little later in the same play, Cranmer declares of her rival's child, the immortal Gloriana,
that—

> " A most unspotted lily shall she pass
> To the ground, and all the world shall mourn her."

In *Titus Andronicus* Lavinia's tears are compared
to

> " The honey-dew
> Upon a gather'd lily almost withered."

The *Schizostylis coccinea* (a colonial from Kaff-
raria) is a most useful and very hardy bulb, bearing,
in late summer, little spikes of crimson flowers which
will go on till frost cuts them down. In Devonshire

they will flower nearly up to Christmas, when the weather is comparatively mild.

We are now not half through the hardy bulbs that deserve mention, but, as it is, the space for indoor-bulbs is dwindling sadly, and we must turn our attention to a few of these.

The bulb of all others that almost every one would like to grow is *Eucharis Amazonica* ; but this is a stove-plant, and cannot be managed in a cool house. (This expression is applied to structures which can be kept in winter at a minumum temperature of 60° in the day, and 50° at night. *Pancratiums* and *Crinums*, again, are both stove-plants; and so, though beautiful and desirable, are not likely to come under the care of the lady of the house, for if there be a stove-house there will be a gardener of some pretensions, and that house will be his private preserve.

The lady need not despair, however. The cool house, in which she gets her way, may be made splendid with amaryllis. Its pretty name is very inappropriate. Amaryllis was the typical country wench of Theocritus and Virgil, and has been associated in modern days with one of Milton's most charming allusions. But the amaryllis of the gardener is a very different person. It is a rich and stately visitor from foreign climes. One species came from South America in the seventeenth century, and so, perhaps, the amaryllis of Milton and of the florist were at least contemporaneous. But a few years back, the only recognised amaryllis was that favourite of cottage window gardeners, the " Scarborough lily"; and now no one who has not seen them can form the least idea of the splendour of the hybrid varieties—scarlet, orange, crimson, rose, and pink, with centred stars, flakes and tips of white, or delicate green; they are a fairy vision of beauty.

The Bermuda, or Easter lily, is a treasure for coolhouse work. *Lachenalias*, too, are so easy to manage, and afford such colour and variety, as to be a very welcome addition to our houses in spring. *Freesia refracta alba* is, however, the nearest to one's heart of those bulbs lately introduced. It is graceful, exquisitely delicate in appearance, and has a perfume like that of cowslips. So long as they are kept out of draughts these Cape bulbs will grow and flower in an ordinary greenhouse. They need no water till they start into growth, but then must have it pretty regularly, and should they, for forcing or other reasons, be moved into a hotter place, the pots must stand in water, for they are water-plants in their native habitat.

Bulbs were for many years the heroines of commerce ; tulips, in particular, being sold for fabulous sums. Holland has long been the headquarters of the trade ; but English gardeners have, of late, been making attempts to grow some of their own bulbs for forcing. The Dutch hyacinth blossoms, which come to the English market in a somewhat drooping condition, are cut off with scythes, and thrown pell-mell into the boxes they travel in. Surely if bulb-culture were more diligently prosecuted here, we should not be so dependent on such ill-cared-for goods as these !

CHAPTER IX.

ROSES ALL THE YEAR.

TO most of us there is a mythical "bower of roses by Bendemeer's stream," in which is enshrined some favourite memory or association. The rose was the theme of Sappho, the plaything of the " wanton boy," and the delight of Cytherea and " the grey old gods."

" O Royal Rose ! " sings Austin Dobson,

> " The Roman dress'd
> His feast with thee ; thy petals press'd
> Augustan brows ; thine odour fine,
> Mix'd with the three-times-mingled wine,
> Lent the long Thracian draught its zest."

It is the flower of all climates and all times. There is scarce a poet who has not sung it, scarce a beauty who has not added its loveliness to her own, scarce a country where one or other species of it does not grow. It is the symbol of England, and at home in the humblest garden and the commonest hedgerow. It has been the type of war, the crown of love; its buds are the emblem of youth, and its dead leaves the expression of enduring virtue; for, like "the action of the just," they retain their fragrance after death. It has been the inspiration of some of the finest lyrics in the English language. Waller sends it to his love, and bids it die—

> " That she
> The common fate of all things rare
> May read in thee—
> How small a part of time they share,
> That are so wondrous sweet and fair."

Herrick bids the rose go bind his love :—

> " Tell her, too, she must not be
> Longer flowing, longer free,
> That so oft has fettered me."

Mrs. Browning, Burns, Hood, and Heine have been amongst the most ardent lovers of the rose. Mrs. Browning's "Song of the Rose," attributed to Sappho, has a jocund strength about it that contrasts agreeably with the pallid sentiment so abundant in the same connection.

> "Ho, the rose breathes of love ! ho, the rose lifts the cup
> To the red lips of Cypris invoked for a guest !
> Ho, the rose, having curled its sweet leaves for the world,
> Takes delight in the motion its petals keep up,
> As they laugh to the wind as it laughs from the west ! "

As for Burns, he is really the Laureate of the rose, for none have written more exquisitely concerning it than he. The rose pulled "wi' lightsome heart" on the banks of Doon, is familiar to all; and the "rosebud by his early walk" has a place in every representative book of "elegant extracts."

Then there is Hood, who writes—

> "Jasmine is sweet, and has many loves,
> And the broom's betroth'd to the bee,—
> But I will plight with the dainty rose,
> For fairest of all is she."

He calls it "Aurora's spright," and says another time—

> "Gone its virgin roses' blushes,
> Warm as when Aurora rushes
> Freshly from the god's embrace,
> With all her shame upon her face."

Heine is full of the rose and the nightingale, treating them with the lightness and distinction peculiarly his own. He often seems to choose the simplest and most threadbare of themes, as if in mockery of the motley crowd who reiterate each other's inanities, and as if he wished to show that what is commonplace in their hands is the clay of the gods in his.

> " Kling hinaus bis an das Haus,
> Wo die Blumen sprïessen.
> Wenn du eine Rose schaust,
> Sag, ich lass' sie grüssen."

Indeed, in every page of song in his volumes he is greeting her. The butterfly is in love with the rose, but whom is the rose in love with ?

> "Ich weiss nicht, in wen die Rose verliebt ;
> Ich aber lieb' euch all' ;
> Rose, Schmetterling, Sonnenstrahl,
> Abendstern und Nachtigall ! "

The spring so stirs him that he wishes he were

> " Eine Nachtigall und sänge
> Diesen Rosen meine Liebe."

He runs over with metaphors, but his love lies hid beneath them. But the world, he says, " nimmts für Poesie." And the world for once is right.

Leigh Hunt says the rose is " the woman of the flowers," and very often in poetry are they linked together.

> " He that sweetest rose will find,
> Must find love's prick and Rosalind."

The hapless Ophelia is the " Rose of May "; while Petruchio, in his mocking courtship, declares his willingness to compare the frowning Kate to

> " Morning roses newly washed with dew."

The roses drooped when Julia was ill; and when
the roses sat in parliament (this pretty absurdity
is very characteristic of Herrick) it was on Julia's
breast—

> " Over the which a state was drawn
> Of tiffany, or cobweb lawn ;
> Then in that Parley all those powers
> Voted the Rose, the queen of flowers ;
> But so as that herself should be
> The maid of honour unto thee."

Tennyson's Lilia is "a rosebud set with little
wilful thorns," and the rose was awake all night for
the sake of Maud. It was an " Eastern rose " the
gardener's daughter was lifting when her lover first
beheld her, and ever after she was "a rose in
roses" to him. " Ah, one rose," he says—

> " One rose, but one, by those fair fingers cull'd,
> Were worth a hundred kisses press'd on lips
> Less exquisite than thine."

Amongst other roses, we cannot forget the fateful
" rose of Yarrow "; nor those two that grew out of
the tombs of Lord William and Lady Margaret
in the *Douglas Tragedy*.

No flower has been more abused than the rose,
no flower has been more superbly, or more inanely,
sung; so that more souls than those of Mr. Cosmo
Monkhouse are " sick of nightingale and rose."

To enumerate the fanciful legends that circle round a rose would be to fill a volume. Every poet has his own pretty fancy, mostly connected with Venus and kisses. There are many ways of accounting for the colour of the red rose. Some say it blushes for shame at having, by its thorn, wounded the foot of Venus as she hastened to help Adonis—

> "My lady's presence makes the roses red
> Because to see her lips they blush for shame."

When Drayton asks the roses " Who with such virtue them inspired ? " they tell him thus—

> "As the base hemlock were we such,
>   The poisoned'st weed that grows,
> Till Cynthia, by her god-like touch
>   Transformed us to the rose."

Carew says—

> "In the white you may discover
>   The paleness of a fainting lover ;
> In the red the flames still feeding
>   On my heart with fresh wounds bleeding."

And of recent poems the prettiest is Mr. Justin McCarthy's " Ballade of Roses," which tells how

> " Venus kissed white roses red."

It seems a little abrupt to turn away from all

this to see what the botanist has to say; but the botanist is a practical person, he ignores verse, despises legend, and describes the *Rosaceæ* or Rose tribe thus :—

" A genus of erect, sarmentose, glabrous, glandu-larpilose shrubs, extending over 'such and such regions.' Flowers ample, showy, solitary, and córym-bose; calyx ebracteolate, tube globose, urceolate, and ventricose; throat constricted, lobes five, very rarely four, spreading leafy, often pinnatisect, deciduous or persistent, imbricated, petals five, rarely four, stamens numerous in series on disc, filaments filaform, carpels indefinite, styles exserted, achenes numerous, included within the baccate tube of the calyx ; leaves alternate, impari-pinnate, very rarely one-foliate, and consisting of connate leafy stipules; leaflets serrated, stipules sheathing at base, and adnate with the petioles." Nothing could be more prosaic !

Thus we have the rose as it appears in verse, in legend, and in science. The true gardener, to whom all these are dear, will find in the rose his greatest delight. The question is, how to secure to ourselves this pleasure every month of the year; and in this short chapter we shall do our best to show that such a result may be obtained.

Of course a judicious selection must be made both for out and indoor growth, and the use (entire or partial) of at least two glass-houses guaranteed. Tea-roses must be largely relied upon, since they blossom at shorter intervals than other kinds, and are in all respects more amenable to cultivation. For outside work they should be budded on the Polyantha or Briar stocks (not Manetti), and for pot-culture on their own roots. May is the earliest time when any roses can be got to blossom naturally out of doors in England; but if that month be fairly warm the hybrid-tea, *Gloire de Dijon*, will flower on south walls, or sheltered spots facing south-west. *Maréchal Niel* also, in a favourable aspect, will put forth the first of its golden blossoms in May; while the humble Scotch briar comes into flower, pink or yellow, and *Rosa Blanda*, a hairy-leaved plant, of North American origin, known also as *R. fraxini-folia*, and *R. Woodsii*. June is called the "month of roses," but in reality only the China, Bourbon, and a few tea-roses will creep out during its earlier half, in addition to those bolder spirits which appeared in May. The last fortnight of June and the first part of July are the weeks when it may be said the air is full of the scent of roses. It needs not to mention any varieties for this season, since almost any kind one fancies can be had in flower now; only perhaps this is a good opportunity to warn the mistress of the rose garden against the eager purchase of new kinds. They may—nay, they are certain to sound charming in advertisements and catalogues; but so often they prove worthless that it is mere folly to buy them to the exclusion of the many old and tried varieties.

Toward the end of July a hiatus will occur, unless thought has been employed to provide for this period beforehand. Some few hybrid perpetuals will probably linger on, and it is possible that tea rose-bushes which blossomed early, and had their flowers cut, will begin again at this period ; but it may also be tided over by the cultivation of several uncommon yet interesting types, such as the climbing prairie rose from North America, known as *R. setigera,* or *R. rubrifolia fenestralis,* also *R. nitida,* another American, flowering in July. It is bright red, and has handsome foliage, turning deep purple in autumn, when the hips are also effective. The Macartney rose generally goes on throughout July into August, also *Rosa sempervirens.* A deep red single rose, common in many parts of Europe, is the *R. rubrifolia,* which comes into flower in August, and may be grown in shrub-

beries, or on banks, etc. *R. Microphylla* is a small
blush rose, a native of China ; it opens in August,
and continues in blossom till October. Then there
is *R. Moschata* (musky) from the southern shores
of Europe, found also in Northern India, bearing
clusters of small pale yellow flowers ; it is a very
pretty climber, flowering in August.

September brings out both tea-roses and hybrids
again. The flowers at this time are not so plentiful
as in June, but they are finer in shape and colour,
and often the more delicately scented ones have
increased fragrance in the dry but comparatively
cool weather of September. The colour will be
perceptible in crimson roses, which are more brilliant now than in early summer; and the teas which
have flashes of pink or rose on their petals will add
brightness to these tints as the weather gets cooler.
It is curious that some few varieties which seldom
flower in perfection in summer are faultless in
shape at this second blossoming. This is almost
invariably the case with *Souvenir de Malmaison*, for
instance. The best flowers we ever saw from this
rose grew on a warm cottage wall—in fact, up the
kitchen chimney, at the beginning of a fine October.
But be October fine or otherwise, the end of the
out-door roses will come during its course, so a few

directions must now be given for the management
of pot-roses to continue the succession.

The plants will, of course, have been standing out
of doors all the summer; except those that climb inside the glass-houses, or that are planted out in beds,
if these exist in the rose-house. It will be best to
keep two distinct batches. They should have been
repotted in spring, after flowering, if it was required, and should be sunk in trenches, if possible,
in summer, with ashes underneath to keep out the
worms. They should not be in too shady a place,
as some sun is needed to ripen the new wood.
The first batch should be brought in early in
September, and placed in a house, warm and light,
to bring them on for November and December
flowering. The second lot may replace early chrysanthemums perhaps in a house vacant in October,
and these should follow the others in January.
The permanent plants (against walls or growing in
beds) will come in flower according to the season at
which they were pruned, and to the temperature of
the house in which they grow. It must be remembered that to force roses in the months from
November to April the houses must be kept at 55°
Fahr. at night ; afterwards 50° will suffice. When
the second batch ceases flowering, about the end of

February, the first, if properly managed, will begin again. When these cease, in April, they may go out at once, if desirable, but the second batch should flower again, and may remain under cover till the end of May or beginning of June. It will be necessary to give these plants regular doses of liquid manure of some kind, varied occasionally by weak soot water, for the sake of the foliage.

Tea and noisette roses may be planted out of doors in April; hybrid perpetuals, briars, China roses, etc., any time in open weather from November to March; but November is far preferable. Hybrids blossom on the old wood, teas and noisettes on the new. Noisettes need scarcely any pruning; teas need old weak sprays cut out in such a manner as to induce strong young growth. Hybrids, on the contrary, must be cut back to strong buds on the old wood, and this is done late in autumn, instead of directly the flowering period is over, as with tea-roses. If our readers are ambitious of growing for show they should remember that specimen blossoms of hybrids must come from young plants, for the reason just given. An old tree of John Hopper, Duke of Edinburgh, or La France, will go on flowering for years; but it will never bear such perfect flowers as in the second year.

LIST OF ROSES.

Hybrid Perpetuals.

Abel Carrière	Dark.
Alfred Colomb	Carmine.
Baroness Rothschild	Pink.
Beauty of Waltham	Rose.
Boule de Neige	White.
Charles Lefèbvre	Deep red.
Countess of Oxford	Bright red.
Duke of Edinburgh	Dark.
Étienne Levet	Carmine.
Gloire Lyonnaise	Yellow.
Her Majesty	Pink.
Jean Cherpin	Plum colour.
John Hopper	Deep rose.
Lord Macaulay	Crimson.
Madame Lacharme	Blush-white.
Merveille de Lyon	White.
Paul Neron	Deep rose.
Reynolds Hole	Dark.
Sénateur Vaisse	Red.
Sir Garnet Wolseley	Carmine.
Ulrich Brunner	Deep red.
Victor Verdier	Rose.

Noisettes.

Céline Forestier	Pale yellow.
Lamarque	White, yellow centre.
Rêve d'Or	Amber.
Wm. Allen Richardson	Apricot to orange.

TEAS.

Adam	Blush-pink.
Catherine Mermet . . .	Flesh.
Comtesse de Nadaillac .	. Flesh to deep apricot.
Devoniensis . . .	Yellowish-white.
Gloire de Dijon . . .	Yellow.
Grace Darling . . .	Cream to pink.
Isabella Sprunt . . .	Pale yellow.
Jean Ducher . . . '.	Yellowish-pink.
Madame de Watteville .	. White to rose.
Madame Falcot . . ' .	. Apricot.
Madame Hoste . . .	Shaded yellow.
Marie van Houtte . .	. Yellow to rose.
Maréchal Niel . . .	Clear yellow.
Niphetos	White to lemon.
Safrano	Apricot.
Souvenir de S. A. Prince .	. White.
Sunset	Apricot to pink.

HYBRID TEAS.

Cheshunt Hybrid . .	. Cérise.
Countess of Pembroke .	. Rose.
Lady May Fitzwilliam .	. Flesh.
Reine Marie Henriette .	. Cérise.

CHAPTER X.

TRUE fern lovers can hardly appreciate the full beauty of their pets without some degree of botanical knowledge; for the wonders of fern-life, construction, and propagation, will, to a thoughtful mind, vie with any romance in interest. This cannot be learnt from the ordinary botanical primers, or handbooks, which allude only to Angiosperms (or flowering plants having their seed vessels covered) and Gymnosperms (flowering plants bearing cones, etc., in which the ovules are exposed), many of them dealing with the first only.

Some more copious work must be studied in order to learn how, at longer or shorter intervals, according to the kind of fern, the exospores split open exposing the endospores, from which are developed the *prothalli*. After certain processes of cell-formation and sub-division, the *antheridium* and *archegonium* grow from the *prothallium*. Within the former spring up spirally coiled *antherigoids*, the *antheridium* is forced open, out fly the *antherigoid* cells, each with a tapering tail, as it were, furnished with *cilia*, or fine hairs. Meanwhile, from a single cell of the prothallium, the archegonium is formed, a bulb-shaped structure having a narrow projecting neck, or canal, lined with a sticky substance which exudes at the mouth. Here the wandering *antherigoids* are captured by their *cilia*, some are sucked down the canal, and reach the *oospore*, or centre cell of the archegonium, which becomes fertilised, and so begins the life-history of another fern.

In managing ferns, whether indoors or out, some reference must be had to the natural habits of the plants. The broad facts are generally recognised that they like shade and moisture ; and acting on such meagre knowledge, people often assume that ferns are the best plants for filling dark corners of their rooms, and if kept standing always in water that the situation is exactly suited to maidenhairs and other beautiful specimens.

It is true that ferns do not like bright sunshine, but they require a certain amount of light ; and their flourishing condition when growing naturally in a shaded place, is often as much due to the fact that the spot is sheltered from cold winds, as to the subdued light. Then, the moist atmosphere in which they delight is in no degree atoned for by a saucer full of dirty, stagnant water, which is as repulsive to the roots as the hot, dry air of a living-room is to the fronds of the plant.

The Filmy Ferns, which are the most moisture-loving of all, get too much dried (as to the fronds) in a glass-house heated by pipes or flues ; but can be grown successfully in a sitting-room under a bell glass, which ensures the constant condensation of moisture in the air.

For the general cultivation of ferns under glass the following points should be remembered. They want light, but not sunshine. Hence a house facing north suits them ; or if in a more sunny position, the roof should be fitted with blinds which can be raised on dull days and drawn down on sunny ones. Again, they like a certain amount of air, but must not be exposed to draughts. Watering must be done judiciously. The soil of pot-ferns should never be let get quite dry, but, on the other hand, no fern will thrive if kept always in a sodden condition ; the soil gets sour, the roots become weak, and eventually die. Most people who know anything about ferns at all, are aware that they are given a larger proportion of drainage when potted, but at the same time may fail to see that this drainage is a mere mockery if they never give it a chance to do its part, and carry off the surplus water when the roots have taken up all that they require.

Very early spring, say about February, is the best time for re-potting ferns. The pots used should be clean and dry, and the soil a mixture of fibrous loam, leaf-mould, peat, and *coarse* sand; equal proportions may be used, and all must be well mixed but *not* broken fine. *Adiantums* do better without the peat, but with a double allowance of leaf-mould.

Ferns growing in glass-houses may (like flowering plants) be roughly divided into stove and greenhouse sections. Tropical ferns require a temperature of 65° to 70° at night in winter. In hot weather, both stove and cool fern-houses must be kept damp by constant wettings of floor and walls, but the cool house will not require this in winter, as the condensation arising from the process of watering will be generally sufficient.

In all cases the moist condition of the air must be produced by wetting the *paths*, etc., not by syringing or turning the hose upon the plants themselves. Outdoor ferns are generally grown in rockeries, or in mounds constructed with roots and tree stumps.

The provision of ample drainage should not be forgotten here any more than in the potting of a single
fern. It is the best plan to hollow out the soil
under the site of a proposed fernery, and form a pit
of several feet in depth, which must be filled with
broken bricks and crocks, so as to allow water to
run away. All rocks, roots, etc., used to form the
sides of the fernery should be very firmly fixed,
and placed in such a manner that the soil intended
to rest in their crevices for the reception of ferns
shall be washed *inward* by rain, not over the edge
leaving the roots bare. For the same reason the
sides must not be built too steeply, and there
must not be projecting pieces of stone near the
top, which will prevent the rain from benefiting
those ferns under its shadow.   There are many
Japanese and North American ferns which can
be grown out of doors with native kinds, but
most of them require a little litter (dried bracken,
for instance) laid over their crowns during the
winter.

The following lists may be useful to any one making
a selection. Those named for stove and greenhouse
culture are some of them new, some uncommon
varieties, as every one knows the more ordinary
kinds :—

9

STOVE FERNS.

Adiantum Mooreii (for baskets).

„ *dolabriforme* (for baskets).

„ *Bausee.*

„ *speciosum.*

„ *lunulatum.*

„ *rhodophyllum.*

„ *tetraphyllum gracile.*

„ *Victoriæ* (dwarf).

Aglaomorpha mayeniana.

Asplenium alatum.

„ *formosum.*

„ *fragrans.*

Cheilanthes elegans.

Davallia fœniculacea.

„ *parvula* (dwarf).

„ *Alpina* (dwarf).

„ *heterophylla.*

„ *aculeata.*

„ *pentaphylla.*

Drynaria coronans.

„ *musæfolia.*

Gleichenia dichotoma.

Goniophlebium glaucophyllum.

Gymnogramma chrysophylla grandiceps (crested gold fern).

Gymnogramma Peruviana.

Lygodium circinatum (climbing).

„ *dichotomum* (climbing).

„ *volubile.*

Niphobolus heteractis.

Nothoclœna sinuata.

Onychium auratum.

Pleopeltis fossa.
Rhiphidopteris peltata.
Pteris tricolor.

GREENHOUSE FERNS.

Adiantum ciliatum (basket).

  „ *gracilum.*
  „ *neo Caledoniæ.*
  „ *Mariesii.*
  „ *palmatum.*
  „ *rubellam.*
  „ *Williamsii.*

Aspidium fragrans.
Asplenium viparium.

  „ *fontanum.*
  „ *Zeylanicum.*

Blechnum corcovadense crispum (dwarf tree).
Cheilanthes Clevelandii.

  „ *gracillima.*

Dictogramma Japonica variegata.
Davillia hemiptera (basket).

  „ *lindleyana.*
  „ *Mariesii cristata.*

Dennstedtia davalloides.
Gleichenia dicarpa.

  „   „   *longipinnata.*
  „ *glaucescens.*
  „ *spelunciæ.*

Gymnogramma ochracea.
Lastrea fragrans.

  „ *prolifica.*
  „ *Richardsii multiflora.*

Lomaria ciliata (tree).

  „ *discolor* (tree).

Lomaria Gilliesii (basket).
Microlepia Hista cristata (basket).
Osmunda Japonica corymbifera.
    ,,   *palustris* (evergreen).
Polystichum viparium.
    ,,   *aristatum variegatum.*
    ,,   *Xipholoides.*
Pteris argyrea.
    ,, *scaberula* (basket).
Strutheopteris Japonica.

Of British ferns it is unnecessary to give a list, but it may be news to some readers that many beautiful hybrids have been obtained of late years, the names of which can be learnt by application to any nurseryman who makes ferns a speciality.

Perhaps the chiefest pleasure of ferns is "the green world they live in." We associate them, somehow, most readily with water, with mountain brooks, with deep ditches and hollows, where, sheltered from the wind and storm, they may unroll their fronds at their leisure.

Some of our prettiest ferns grow in just such woodland scenes as this; as, for instance, the male fern, and the graceful and slender beech-fern, the broader though similar frond of the oak-fern, and the lady-fern, which—

> " Uncoiling in the shade,
> Turns serpent folds to plumes of waving green."

What is more delightful on a fresh and dewy May morning than to hunt along the sheltered bank of a mountain brook for tiny young ferns that grow in the moss and moisture there? The power of such a memory is attested by Coleridge, who, in the midst of the description of the lonely ship and the wide ocean, sings—

> "It ceased, yet still the sails made on
> A pleasant noise till noon ;
> A noise as of a hidden brook
> In the leafy month of June,
> That to the sleeping woods all night
> Singeth a quiet tune."

Those whose gardens do not possess such nooks of themselves, should seek to supply them as soon as possible. There was an old superstition that they who gathered the seed of the fern on midsummer eve became gifted with the faculty of invisibility, and they who possess or obtain for themselves an out-of-doors fernery such as this may, after a fashion, realise the old superstition, and have ever a quiet green corner where they may sit and dream unseen. Shakespeare refers in his own shrewd way to this old belief—

> " 'We have the receipt of Fern-seed—we walk invisible.'
> 'Now, by my faith, I think you are more beholden to the
> night than to Fern-seed for your walking invisible.' "

But not all our ferns love the woods. The bracken, for instance, has no such retiring habit, but is the glory of the mountain-side. It is a familiar item in Scotch hill-pictures. Burns made love

> "Among the brackens on the brae,"

and again refers lovingly to another trysting-place in

> " The lone glen o' green bracken,
> Wi' the burn stealing under the lang yellow broom."

Sir Walter Scott talks of the bracken as the High-lander's pillow, and the author of "Olrig Grange" does not forget it. A northern hill when the bracken clothing its slopes has faded to golden-brown is not readily forgot; and any one who travels by the Scotch express through the hills of the south of Lanarkshire in October will see this to perfection. It is near here that the old home of the Douglases lies; and "the good Sir James," when he died fighting in defence of the Bruce's heart in Spain, requested—so the old ballad runs—to be taken home and buried amid his native bracken.

The Royal Fern is one that, though its native home is in the bleak, open bogland, shows to most advantage when grown indoors. It is a savage which yields readily to the softening influences of

civilisation.   Its leaves, though naturally rough and
dark, grow up green and delicate as any tropical relation.   It will stand any amount of transplanting
and knocking about, and is the most good-
humoured and accommodating of royalties.

CHAPTER XI.

ORCHIDS.

THE day has gone by when it was thought that orchid culture was possible only for the very rich. Small orchid houses are included in quite modest establishments, and some popular varieties have been found to do well in " mixed houses."

On the other hand, the specialists and orchid fanciers never expended larger sums on their favourites than at the present time. Occasionally, one reads scathing remarks on the folly or iniquity of this Duke or Duchess, that Baron or M.P., because they have enlarged their orchid houses, or given a long sum for some newly imported specimen. Yet if with one consent these wealthy people decided to give up orchids, and devote the money to charity, the first claimants on their bounty would be a procession of those thrown out of employment by such a decision, beginning with the collector who ransacks the tropics for choice varieties, the importer

and all his staff, and concluding with the lad fresh
from the village school who stokes the fire and
washes the pots in the ducal gardens.

In the chapter on Ferns it was remarked that
a fern-grower should, in order to appreciate her
subjects, make herself acquainted with their structure
and mode of fertilisation. This is even more urgent
in the case of orchids. No one can do better than
study Darwin's book " On the Various Contrivances
by which British and Foreign Orchids are Fertilised."
Botanists recognise 334 genera, and about 5,000
species. Broadly they divide the genera into five
tribes—the *Epidendreæ*, containing eighty-eight
genera ; the *Vandeæ*, a hundred and one ; *Neotteæ*,
eighty-one ; *Arphrydeæ*, thirty-two ; *Cyprepedieæ*,
four.

Gardeners, for practical purposes, divide this
great wealth of plants very simply under three
heads, according to the temperature in which they
are found to thrive, when under cultivation. These
are *Hot*, *Intermediate*, and *Cool*, or sometimes *East
Indian*, *Brazilian and Mexican*, and *Peruvian*.

Those in the first class come from the mid-tropics
of either hemisphere (the expression East Indian
being merely technical), and require not only heat
but a very moist atmosphere at the period when

they are in active growth. The houses in which they are placed must not be suffered to fall below 75° in the daytime during winter, or 65° at night. The Brazilian and Mexican, or intermediate orchids, do not require nearly so much moisture; but on the other hand, they shrivel if the air is quite dry. Sixty degrees must be the lowest temperature on winter nights. The cool orchids are those found in mountain regions, chiefly in South America and India. Many of these will flower almost all the year round, and therefore they are special favourites with amateur growers, who should not grudge keeping up the mild, damp atmosphere beloved of these plants. Even on a summer day they do not require more than 70° of heat, and a fall to 40° on a winter night will not injure them.

As to the general management of orchids, it may be remarked that, next to warmth and moisture, cleanliness is perhaps the most important matter. Every part of the plants, their soil (or what serves for soil), pots or baskets, and all surroundings, must be scrupulously clean. When planted in pots these must be *half-filled* with clean drainage, above which is used Sphagnum moss, and fibrous peat mixed with charcoal. Pots or pans should be no bigger than is absolutely necessary to pack the roots in. Shallow pans for hanging, or baskets of teak-wood, are better suited for most varieties.

It must be understood that hot orchids must have a house to themselves, and the best arrangement for securing the required moisture is to have a water tank in the centre, underneath the stage; a pipe from the roof should supply the tank with rain-water. Watering with *soft* water is very important, and should always be done where possible. Orchids never ought to be fumigated to destroy insect foes. Careful sponging is generally enough, and if more is needed, the leaves should be washed with soft soap.

Naturally, the hot orchid house is that requiring most care, for a stagnant condition of the air is most unfavourable, therefore ventilation is needed, but must be contrived so as not to reduce the temperature too suddenly, nor yet to let cold currents of air touch the plants. The ventilators in such a house must be placed low, on a level with the pipes of the heating apparatus. Again, the hot orchids nearly all have their periods of rest, during which time water must be very judiciously administered.

It is, of course, impossible to give names of many varieties here, but the following are some of the most interesting :—

Hot or " East Indian " Orchids.—*Peristeria elata*, named *Spirito Santo* by the Spaniards, who first saw it at Panama, from the resemblance of the white petals to a dove taking flight. Then there is the beautiful, but terribly expensive, *Vanda Sanderiana*, from the Philippine Islands. Fortunately the Javanese varieties are very beautiful, and much more obtainable. One of the loveliest is *Vanda teres*, which loves to climb on the roof in the hottest part of the house. Another with much the same proclivities is *Renathera coccinea*. *Cattleya superba* and *Dowiana* are two of the most beautiful occupants of an orchid stove ; *Cattleya guttata Leopoldii Acklendiæ* (dwarf) and *Camethystoglossa* like *plenty* of sunshine, for lack of which many growers give up *C. superba* as a hopeless case.

On the other hand, the *Phalænopses* (natives of the Philippines, Java, Borneo, Cochin China, and Burmah), need careful shading, as do the *Angræcums, Ærides*, and stove *Cyprepediums*. *Cælogyne pandurata, Dayana, Massangeana*, and *Sanderiana*, are all " hot," new, and expensive; and connoisseurs tell us that the exquisite *C. cristata*, though rightly treated as an " intermediate " orchid, yet does better if brought into the hottest house, when it first starts into growth. It grows, in its natural habitat, in very hot and exposed situations. There are many beautiful *Epidendrums* requiring hot treatment, which are easy to manage, and sufficiently cheap. Their names may be gleaned from any catalogue. The same may be said of stove *Dendrobiums*, hailing mostly from Australia, of the *Lælias*, and a few *Oncidiums*.

The greater number of *Cattleyas* and *Lælias* will flourish in the warm or intermediate house, which is indeed known in some establishments as the *Cattleya* house. A part of this house should be fairly shady, to suit *Bolleas*, *Cœlogynes*, warm *Odentoglossums*, and *Masdevallias*. *Lælias* and *Cattleyas* do best if very thoroughly supplied with water, and then left a week or ten days without. The intermediate house should never be syringed.

Warm orchids mostly come from sub-tropical climes, or the highlands of very hot countries. One of them, much sought after at present, is *Odontoglossum vexillarium*, now procured easily enough, but such a rarity five-and-twenty years back, that a single flower was "lent," as a great favour, to a botanist, on condition he would show it to no one else, nor draw, nor photograph it. Of *Cattleyas*, it is said there are sixty species grown, and some of these have twelve, twenty, thirty, even forty varieties. Two of the species have the distinction of owning private and particular insect pests, all to themselves : the rare and beautiful *C. aurea* has a persecuting beetle in its track, and *C. Mendellii* a fly which lays eggs in the very centre of the young shoots. These are imported with the orchids while themselves in the egg form, but happily they soon succumb to the English climate.

Some of the *Dendrobiums* are delightful to grow, by reason of their ability to flower in winter, if only the grower can get them well started in early summer. These are *Dendrobium Wardianum, Falconeri, Nobile, Devonianum, Crassinode,* and *Crystallinum.* There are several "warm" *Angræcums,* most of them natives of Madagascar. One is said to be a native of Japan, but it is believed very unlikely; and the chances are that the Japanese imported and naturalised it long ago.

The cool orchids are, naturally, the easiest of cultivation, and the most likely to be attempted by a lady who actively participates in the care of her own houses. An eastern aspect is very suitable for these plants, as it is a light one for winter, while in summer the sun is off before it becomes too bright and strong. Houses with west or even south aspects can be used, if carefully shaded, in such a manner as not to exclude light or air. A canvas blind should be used, supported by strips of wood, raising it several inches above the glass, so that when, during hours of sunshine, it is necessary to let down the blind, a current of air can pass between it and the glass. Whether by sun or fire, the temperature should never rise above 65°. There must be top and bottom ventilators to such a house ; the lower ones may be opened a little every day when the outside thermometer registers over 40°. Top ones will be used only in summer ; and not then unless it is warm, wet weather. On hot, dry days these must be kept shut, lest the inside moisture should escape. The floors must be kept damp by constant sprinkling. If this is attended to, the plants will require to be watered only about once in five days.

The following is a good selection for any one wishing to make a first start in cool orchid growing. Be it remarked that selection is not easy for the beginner, as it is calculated that there are at least 2,000 " cool " varieties. Many small growers—by which term nothing opprobrious is meant, only that their space is limited—confine themselves entirely to *Odontoglossums*, *Oncidiums*, *Cypripediums*, and

Lycastes. Such people are very wise ; but wisdom comes of experience, and amateurs usually desire plenty of variety, so other genera are mentioned here :—

> *Ada aurantiaca.*
> *Cattleya citrina.*
> *Cœlogyne barbata.*
> „   *cristata.*
> „   *speciosa.*
> *Cymbidium Lowii.*
> *Cypripedium Harrisianum.*
> „   *insigne.*
> „   *candidum.*
> „   *parviflorum.*
> „   *pubescens.*
> „   *macranthum.*
> „   *venustum.*
> „   *Sedeni.*
> *Dendrobium infundibulum.*
> „   *Jamesianum.*
> *Disa grandiflorum.*
> *Lœlia autumnalis.*
> „   *Dayana.*
> „   *majalis.*
> „   *præstans.*
> *Lycaste aromatica.*
> „   *cruenta.*
> „   *Deppei.*
> „   *Harrisonii.*
> „   *Skinneri.*
> *Masdevallia bella.*
> „   *Chimœra.*
> „   *infracta.*

Masdevallia rosea.
Maxillaria grandiflora.
Odontoglossum crispum.
         „   *citrosmum.*
         „   *gloriosum*
         „   *hastilabium.*
         „   *odoratum.*
         „   *phalœnopsis.*
         „   *vexillarium.*
Oncidium crispum.
         „   *cucullatum.*
         „   *Forbesii.*
         „   *ornithorhynchum.*
         „   *macranthum.*
Pilumna nobilis.
Sophronitis grandiflora.
         „   *violacea.*

The *Odontoglossums* come to us from America, where many a life has been lost, and really terrible hardships undergone, by the collectors who travel the mountain regions of Columbia in search of new kinds, or fresh stocks of valuable ones. Many of the beautiful and brilliant *Masdevallias* inhabit the same regions, and for some reason or other they are found to stand transportation very badly. It is said that once cases were sent home containing 27,000 plants, of which only two were alive on reaching England.

Lycastes also come from South America. They are denizens of the woods and forests of Guatemala.

10

Cypripediums are found in almost every quarter of the globe. We have one native variety (*C. calceolus*). It is strange that many growing naturally in decidedly cold places will not stand the damp of English winters, however well protected.

The *Oncidiums*, again, are American, and grow in lofty positions ; they generally flourish best in the coolest, nay, *coldest* part of the house.

The poet who shall sing the orchids has yet to be born ; his time can scarcely arrive until more euphonious, if more familiar, names, have been given to these lovely flowers. *Odontoglossum* or *Dendrobium*, and other of the crack-jaw titles conjured up by botanists, make all but comic rhymes hopeless, and rhythm impossible.

Our own native orchids seem insignificant when compared with those of the tropics ; but their delicate beauty well repays study, and some of them should be found among the flower-besprent turf of a wild garden. They have mostly descriptive English names—the Bee, the Monkey, the Lizard, the Butterfly, Orchis, Ladies' Tresses, and Birds' Nest. Of folk-names we are acquainted with one only, which is scarcely poetic. It is given in the Midland Counties to the early purple *Orchis mascula*, which is there called " Bloody Butcher."

The economic value of the Orchid tribe is not great. A kind of glue is made from some of them, and a nutritious substance called Salep used to be prepared from the *Orchis mascula*, and the popular flavouring essence of the *Vanilla aromatica* is the only product of the whole 3,000 species that may be said to form an article of commerce.

As we said before, there is much that is romantic in the records of orchid collecting, and not unfrequently the romance deepens into tragedy. At the time of writing, a great sale has been held in Cheapside of 1,000 plants of *Dendrobium phalænopsis var. Schröclerianum*. The first consignment of 400 was burnt with the vessel in which it was shipped from New Guinea. The collector telegraphed for instructions, and was bidden to go back and try again. Returning, he found a perfect mass of the glorious flowers luxuriating among the bones and skulls of the burial place of a savage tribe. After much persuasion, and bribes of beads, mirrors, brass wire, and coloured handkerchiefs, the people permitted the desecration of their ancestral tombs, and so far entered into the spirit of the thing as to perform a war-dance around the packing-cases when ready for the start. Only they insisted, as a necessary precaution against vengeful gods or departed heroes, that the " Idol of the Golden Eyes " should go with the weeds—as they consider the plants to be. This precaution was less absurd from their point of view than appears at first sight, for one of the finest plants was brought over, exactly as found, growing out of a human jawbone.

CHAPTER XII.

THE CONSERVATORY.

THE practice of growing flowers and fruit under glass is a comparatively modern one. The wealthy citizens of Rome tried to shelter their more precious plants, but the only material available for them was thin layers of transparent stone.

It was not till the fifteenth or sixteenth centuries that glass was used for this purpose at all, and the protection of orange-trees was the necessity that called for the invention. The first glass-house to which we can affix a date was that erected in 1619 by Solomon de Cans, of Heidelberg. In the early days of greenhouses glass was used only for the walls; the glass roof was an afterthought of the eighteenth century, and the arched or curvilinear roof an artifice of the nineteenth. The method of heating was at first primitive in the extreme—hot embers being placed in a hole in the floor ! The present elaborate arrangements of stoves, or boilers and pipes, heating whole ranges of houses, developed gradually out of these somewhat impotent beginnings.

Glass-houses were a rare luxury, even in the magnificent gardens laid out in France during the last century. Neither Versailles nor Marly in its first magnificence could boast of a conservatory. General Resson brought home a coffee-plant; but in the chill northern climate it soon died. The burgomaster of Amsterdam presented another to the king; and experience having taught them to utilise glass, the plant flourished. King Louis delighted in growing his own coffee, and roasting and grinding his own beans. It was a mighty favour to partake of the king's special coffee; and discreet courtiers declared it equalled the best colonial product. Even they did not say it surpassed it.

Tall houses were not thought of up to the time when *Cereus Peruvianum*, brought home as a small plant, kept growing taller and taller, till its house could not contain it. Its glass roof was heightened gradually, until at last it reigned alone in a tall glass-house like a belfry !

This term conservatory, as every one knows, is very elastic. It may mean the highly ornate erection opening out from the drawing-room or boudoir of a splendid mansion, carefully planned by the architect so as to be in perfect congruity with his designs. On the other hand, it is bestowed on many a humble lean-to, added at a low cost to an unpretending suburban villa, or a mere glazed cupboard projecting from a London lobby.

There are many owners who would scorn to talk of their greenhouse, but derive great satisfaction from the mention of their conservatory. It is such a harmless source of pride, that on no account would we interfere with it ; we merely allude to it since the wide application of the word makes some kind of definition necessary on our part. Be it understood, then, that in this chapter, which for convenience' sake is headed " Conservatory," all such glass structures are alluded to as are known by horticulturists as cool houses. That is, those which, when artificially warmed, are kept at a temperature not falling below 50° at night during winter. It may fall several degrees lower during very severe weather for a few hours, without permanent injury to the plants, but 50° should be the average night temperature during winter.

This is the kind of house most likely to fall under the charge of a lady, and she will not find it very onerous if the conservatory is merely replenished from a range of other houses, where the stages of development and of shabbiness are gone through, the plants being removed into the conservatory only while at their best. Attention to watering, ventilation, and cleanliness is all that will then be required. But if the house is a general one, perhaps supplemented by a few frames only, then a considerable amount of care and trouble is necessary to make it attractive. A few practical rules will, in either of these cases, be helpful to the presiding goddess of the conservatory. First as to watering, the great standing difficulty of amateur gardeners.

Call to mind pictures of ladies engaged in gardening, and it will strike you how strange it is they are always watering, and that their water-pot invariably has a rose upon it. A rose seems to be popularly considered a feminine adjunct, whereas it is really far more often required by the practical worker in the propagating-house, or on the seed beds. There is one case, however, in which the rose should be used—that is, in watering newly potted plants, which should be thoroughly soaked in this manner, as the water falling in a shower washes the soil home among the roots and fibres. After this soak the plants should not be watered much until the roots grow out towards the sides of the pots; because this unoccupied soil, if kept wet, becomes unwholesome, the roots will not strike into it, and the plant will presently die. Plants want most water at the time they are growing fast, throwing out new leaves, buds, and flowers. On this and on the temperature the amount of watering depends. A fuchsia in full blossom in July may need soaking twice a day; in September every other day may be sufficient for the same plant, and through the winter it may need none for weeks, or months. No one time of watering will apply to all the plants in one house. Some will have taken up much more moisture than others, either from their nature, or temporary needs; or even from their different position on the shelves. Small pots get dry sooner than large ones containing a mass of soil; and with those standing alone, or in places where all sides of the pot are exposed, evaporation takes place much sooner than with pots set close together.

Finally, watering should not be delayed until the plant flags for want of it, neither should it be given before the soil is *rather* dry. The wants of each plant must be considered individually; it is no use to give them all stated portions at stated times, like the inmates of a workhouse.

In fine, mild weather give as much air as possible, only avoiding draughts. During summer it will often be advantageous to have a little air on all night as well as in the day; but in early spring and autumn only the best hours of the day, say from 11 A.M. to 3.30 P.M., are suitable, and a few mid-day hours, on dry, mild days in winter, will suffice.

Pains must be taken to keep the temperature as even as possible. Very sudden changes from heat to cold are very hurtful, and will cause attacks of greenfly, if nothing worse. It is wiser to keep a strictly moderate temperature, varying in winter only 10° between day and night, than to get the heat up much higher every now and then, and at other times to let it drop below the average.

Space scarcely admits of fuller directions being given here, but many of the remarks in the two following chapters will apply to the general management of plants in cool houses.

Plants suitable for growing in conservatories are roughly divided into four groups: soft-wooded, hard-wooded, bulbs, and climbers. Soft-wooded plants are such as have soft stems with much sap, and are mostly propagated from seed, or by cuttings. They do not thrive in a close atmosphere, and dislike extremes of sunshine or of shade. In the first they droop, in shade they will not flower.

The most popular soft-wooded plants are pelargoniums, brilliant colonials from the Cape of Good Hope; calceolarias or slipperwort (*calceola* being a slipper, although some say it took its name from Calceari, an Italian botanist), familiar to all; balsams, fuchsias, *Eupatorium*, salvias, cinerarias, primulas, and spiræas. Balsam is said by some to be the original of the "moly" of the Ancients, although others consider it to have been one of the foreign garlics. Balsam, so one story runs, was first seen growing in the vales and low grounds of Peru, where the Indians worshipped it, and called it Molle, as being soft and gentle to the touch. From Peru it was brought to the king's garden at Madrid. Both Ovid and Homer refer to the plant—

> " Moly the gods would have it named.
> It's held to have a root that's black as night."

It was moly that Hermes gave Ulysses to enable him to withstand the enchantments of Circe.

> " Black was the root, but milky-white the flower."

And it is recorded that

> " On th' all-bearing earth unmarked it grew."

Dioscorides, however, said that it was a kind of garlic. And as for Gerarde, he scorns the whole thing, and says : " As for repeating of foolish and vaine figments, the conjuring of witches and magicians, enchantments which have been attributed unto these herbs, I leave them to such as had rather plaie with shadowes, than bestow their wits about profitable and serious matter."

The fuchsia has rather an interesting history. It is said to have been brought first to this country by a sailor, who placed it in his cottage window. Mr. James Lee, a nursery gardener at Hammersmith, saw it there, noted its novelty, and went in and bought it. It originally came from America, and though common in Central and South America, made itself at home in Yankeeland as well. Mr. Aldrich has some charming verses about it—

" The chestnuts shine through the cloven rind,
  And the woodland leaves are red, my dear ;
The scarlet fuchsias burn in the wind—
  Funeral plumes for the year !
The year which has brought me so much woe,
  That if it were not for you, my dear,
I could wish the fuchsia's fire might grow
  For me as well this year."

The plant was called after Leonard Fuchs, a German botanist of the sixteenth century. It belongs to the same family as the evening primrose, which, by the way, is no relation to the pretty little sister of the cowslip. The coloured calyx of the fuchsia has been diversified from its original crimson to an astonishing variety of shades.

The fuchsia is, however, but an upstart *parvenu* compared to the *Eupatorium*. The latter is of ancient and honourable lineage. Dioscorides described it ; and Pliny relates that it owes its name to Eupator, King of Pontus, who discovered in it a valuable antidote to certain poisons. There was no more useful defender of monarchy in those days than an antidote. The salvias are from Mexico, and are, in fact, the gay and useless elder sisters of that fragrant Cinderella, our English sage (*Salvia verbenaca*). Primulas and spiræas have also their English connections familiar to all, and as beautiful in their wild woodland grace as any.

Hard-wooded plants want a closer atmosphere than the preceding, especially after flowering, to encourage new growth. When this is formed they may stand out of doors through the summer in most cases. They must never be allowed to get thoroughly dry at any time, or they will lose their leaves ; whereas such of the soft-wooded plants as are not annually grown from seed are the better for being kept dry through the winter. Many of the hard-wooded plants are evergreens. Some of the most familiar are acacias, azaleas, *Boronias, Billobas, Bouvardias,* heaths, camellias, *Abutilons, Epacris, Cytisus,* etc.

The acacia is delightfully decorative, and has the merit of belonging to a singularly useful family. One of the species yields gum arabic, and others have oleaginous and saponaceous merits. They are cousins of the famed "Sensitive Plant," and the acacia was one of the favoured nine (the number of the Muses) who waited in the garden for Maud. The graceful aerial ways of the plant have caused it generally to be described as in motion. Moore talks of the acacia waving her yellow hair. Tennyson says—

> "The slender acacia would not shake
> One long milk-bloom on the tree."

And Swinburne speaks of

> "Rose-mouthed acacias that laugh as they climb,
> Like plumes for a queen's hands fashioned to fan her
> With wind more soft than a wild dove's wing."

The azalea is especially charming because of its infinite variety of colour.

> "Azaleas flush the island floors,
> And the tints of heaven reply."

The honey of the azalea contains a narcotic poison, and Xenophon's soldiers having partaken of it when retreating from Asia, were overpowered by stupe-faction and delirium. It is strange to think that magnificent plants like azaleas and rhododendrons should be related to the minute and dainty heath of the greenhouse, and the wild and ragged heather of the mountains and moorlands.

Boronias are charming on account of their scent; but *Boronia alata*, although it has a pretty flower, should be avoided, as the perfume of its leaves, once bruised, is rather too pertinacious to be pleasant. The *Boronias* derive their name from Francis Boroni, the Italian servant of Dr. Sibthorpe, the author of the *Flora Græces*. Boroni was a very valuable aid to his master, and owed his death to an accident at Athens. Surely his appreciative master could have chosen no better method of retaining the memory of his name than this. The *Boronias* belong to the same family as the common rue, whose curious lingering smell we all know.

Camellias are familiar plants. Mr. Ellwanger derides their waxy rosette-like flower. They are curiously characteristic of the spirit of their native land; they chiefly come from China. There is no " sweet disorder" about them, they have a very correct, self-satisfied air when well grown. They owe their name to Camellus, a Moravian Jesuit, who travelled in Asia, and wrote a history of the plants of the Isle of Luzon.

The *Abutilon* is a relation of the familiar marshmallow; and the *Cytisus* is better known to some as the broom (*Genista*), whose pretty branches of blossoms, " yellow and bright as bullion unalloyed," at once proclaim its relationship to the long yellow broom of the hillside and the laburnum— " dropping wells of fire," as Tennyson calls it; " golden rain " as the Germans term it.

Bulbous plants, especially Cape bulbs and Japanese lilies, are among the most ornamental things that can be used for conservatory ornamentation where space is limited, in that they all need seasons of rest, and at such times may be stored in any cool, dry place, so that they leave the shelves and stages free for other plants. The same merits appertain to the beautiful tuberous begonias, surely the most welcome of all late innovations of the kind. Among the true bulbs and corms suitable for conservatory use are *Babianas*, the *Achimenes*, *Cyclamen*, *Freesias*, *Crinums*, *Nerines*, *Amaryllidaceæ*, *Gladioli*, and *Lachenalias*.

Cool-house climbers are almost too numerous to

mention. Tea-roses and noisettes should always be
admitted; and others easy to manage are, *Bougainvilleas, Plumbagos, Passifloras,* and their close
connections the *Tacsonias,* fuchsias, some of the
more delicate clematis, *Cobea scandens, Ficus repens,*
tuberous tropæolums, etc. These last are very
desirable, as their gay flowers will go on all through
the winter, making a brilliant spot of colour amid
the neutral-tinted flowers and sombre evergreens of
the decadent season.

CHAPTER XIII.

THE HOTHOUSE.

BY this is meant a glass structure kept at a high temperature, suitable to the growth of tropical or semi-tropical plants, and the propagation of delicate species from seed, or by cuttings. Ordinarily the temperature of such a house will vary from 70° to 80°; and in most cases it is advisable to keep the air very moist. Some persons have an impression that a dim light is requisite in a stove-house, but this is a mistake; for though bright sunshine pouring through the glass is calculated to do mischief, and should be guarded against by blinds, yet a good deal of light is necessary to the plants, particularly those whose value depends on their tinted leaves, such as crotons or caladiums. Flowers, again, will be neither so fine, nor so numerous, when the roof is covered either with artificial shading, or by thick growth of creepers.

As stove-plants require a great deal of water, it is specially needful that a large proportion of drainage be used in potting them, and that the soils be light and open, so as not to get clogged with water.

The moisture and heat are, unfortunately, favourable not to plant-life only, but also to that of all kinds of insect-pests; so that if cleanliness is important in all departments of gardening, it is absolutely indispensable in the management of a stove. Floors, walls, roof, stages, plants, pots, and the very sticks and ties must be free from dirt and dust. No kind of litter should be left about even in a cool house; but under the influence of heat and moisture it decays much more quickly, and produces mould, harbours insects, and in other ways works ruin.

It must further be remembered that in a stove-temperature plants grow very rapidly, and should never be crowded closely, or they get drawn and "leggy," or one-sided (according as the light falls), in a very short space of time.

Constant syringing will keep leaves clean in an ordinary way, but it cannot be applied while there are plants in blossom, else the petals get spotted and spoilt. Moisture must be secured in this case by constant sprinkling of the floors, and individual plants kept clean by careful hand-washings with soft water, a sponge, and soft soap if necessary. Some stove climbers are very liable to get dirty and infested with insects ; this is especially the case with stephanotis, which should be loosened entirely from the wall or trellis on which it grows, and cleaned all over, without disturbing the root. Fallen or withered leaves, old sticks, and ties removed from a plant infected with scale, mealy-bug, etc., should all be burnt at once.

The most treasured occupants of the hothouses are orchids ; but as they have been dealt with separately, they may be passed over here.

Of the better-known flowering plants we may mention :—

Anthuriums.	Eucharis Amazonicas.
Allamandas.	Imatophyllums.
Gloxinias.	Epiphyllums.
Gardenias.	Euphorbias.
Ixoras.	Pancratiums.
Francisceas.	Poinsettias.

Dr. Allamand first gave the seeds of the *Allamandas* to Linnæus. They are the tropical cousins of our own charming periwinkles, and their curious stigmas are worth noting. Gloxinias and *Pancratiums* are both tropical bulbous plants; whilst the

Euphorbias and the handsome scarlet flowers of *Poinsettias* belong to the great *Euphorbia* order. The gardenia, curiously enough, is a member of the quinine-producing family. The *Cinchonaceæ* is indeed an order of very varied distinction. The fruit of one plant is sold in China and Japan for dying silks yellow, and the wood of some of the gardenias (*G. Humbergia* and *G. Rothmanni*), is so hard that it is used at the Cape of Good Hope for making agricultural implements. Nor is quinine the only important product of this family; both ipecacuanha and coffee are included in it. The *Anthurium* is welcome on account of its huge shield-like leaves, with their fine variation, as well as because of its curious red flower. Of *Ixora* as it grows wild Miss Gordon-Cumming gives a good description. After a season of great drought in Ceylon, when most of the plants were shrivelled and spoiled, she says : "The scentless scarlet *Ixora* alone survived," and in such abundance as to be tiresome; especially as "colonies of vicious red ants made their home among its blossoms."

Of foliage plants we may mention

Aralias (the choice varieties).   Crotons.
Foliage begonias.   Dracænas.
Asparagus plumosus.   Caladiums.
Pandanus (*vars.*).

And numerous members of the palm families, such
as—

> *Areca lutescens.*
> „ *Madagascariensis.*
> „ *rubra.*
> „ *speciosa.*
> *Calamus asperrimus.*
> „ *ciliaris.*
> „ *spectabilis.*
> *Chamœdorea elegans and gracilis.*
> *Cocos Weddeliana.*
> „ *flexuosis.*
> *Thrinax elegans and radiata.*
> *Veitchia Canterburyana.*

The aralias are interesting on two grounds, beside
that of beauty; one, that they are the foreign
relations of our English ivy, and the other, that
the cellular tissue of *A. papyrifora* supplies the
Chinese with the rice-paper on which they paint
such gay, delightful birds and curious quiescent
butterflies. Of *Asparagus plumosus* one cannot
have too much, for nothing equals the decorative
effect of its seaweed- like foliage. *Crotons* and
Caladiums are rich in colour. Miss Gordon-
Cumming tells how the latter befringed the edge
of the verandah of a friend's house in Ceylon. Of
this lady's latest work ("Two Happy Years in
Ceylon") not the least interesting portions are those

in which she describes " Hothouse Plants at Home." We have often lingered in a hothouse, and wondered how all the heterogeneous mass of bloom imprisoned there looked when on its native soil. One friend has told us of the crotons growing wild, and free, and strong, in the Indies ; and here and there we have obtained glimpses of the charming home-life of our exotic acquaintances.

> " And the red passion-flower to the cliff,
> And the dark blue clematis clung.
> And, starred with myriad blossoms,
> The long convolvulus hung."

Does not the glass feel strangely cool and clammy to the passion-flower, accustomed to rougher bark or cliff ? And how dull and poor and limited must the uninteresting area of the average hothouse seem after the glorious verdure and gorgeous colour of its home ! Look at the areca palms, growing in a trim greenhouse; then read Miss Gordon-Cumming's account of " a group of stately areca palms, faultlessly upright, like slender alabaster pillars in a leafy sanctuary, each crowned with such a capital of glossy green as human architect never devised."

She tells of a native proverb which runs : " He who can find a straight cocoa-palm, a crooked areca, or a white crow, shall never die."

The *Dracæna*, too, is a stately plant in its native haunts, and there is one in the Canary Islands which is seventy feet in circumference at its base! The *Dracæna* belongs to the lily family, and is, comparatively speaking, dumpy—ten or fifteen feet being its average height. It has the merit of being easily propagated. *Calamus* is the name of the reed pen employed by the ancients, and *Pandanus* owes its name, screw-pine, to the cork-screw method in which the leaves arrange themselves on its stem. Some of the stunted palms we see in small hothouses are very pitiful. The palm has come to the land of the pine; yet, strange to say, the union is not attended by any very brilliant results. We remember still the bewildered expressions, half of incredulity, half of mirth, of two West Indians, seated opposite a " table palm " at a friend's house. They evidently had difficulty in recognising their stately eastern friend in these debased conditions.

Some of the best stove-creepers are,—

Bougainvilleas.	Jasminums.
Bignonias.	Passiflora.
Clerodendrons.	Smilax macrophylla.
Dipladenias.	Stephanotis.
Hoyas.	Thunbergia Harrisii.
Ipomœas.	

The *Bougainvilleas* owe their name to the first

Frenchman to voyage round the world (1766-69). The *Bignonia* is, of course, the trumpet-flower, and the *Dipladenias* and *Hoyas* are latescent plants belonging to closely allied orders. The *Ipomœa* is the maritime convolvulus, and this is how it grows in Ceylon. " The shore was carpeted with goats' feet, ipomœa, a large lilac convolvulus, whose glossy green foliage, with a profusion of delicate blossoms, mats the sands to the very brink of the sea, affording shelter to thousands of tiny crabs. This pretty plant flourishes on the sea-board of all parts of the isle, and constitutes one of the many charms of the beach." And another time we have a passing glimpse of " a lilac and green carpet of convolvulus, beneath a grove of tall graceful cocoa-palms bending in every direction."

The *Thunbergia* seemed equally at home in Ceylon, although it was only a naturalised foreigner, and came across the sea from Burmah. " The exquisite thunbergia, starred with myriads of blue-grey blossoms, climbs from a carpet of the freshest, richest grass to the very summit of a large group of trees, thence drooping in graceful festoons, and linking them altogether into one fairy-like sanctuary, haunted by dainty birds and radiant butterflies. I always remember the sunlight falling through that exquisite veil of delicate grey and lavender as an ideal of tropical perfection."

Does the *Thunbergia* ever look quite like that in our hothouses ?

It is not given to every one to reproduce the tropics fitly under glass. It requires a vast amount of money and ingenuity ; and when it is done, it is, after all, only a hint at the original. Still, if it suggest thoughts of the wonderful countries that have been the inspiration of so many, it is something.

CHAPTER XIV.

MODERN writers often tell us, as if with a sigh, that this is the age of prose and common-sense; but as one looks round, and listens to the hurly-burly, the parrot cries of favourite teachers caught up by their pupils, it seems to be undoubtedly the age of fads and cranks. And many of the fads are so purely sentimental, that they form a striking contrast to the prosaic tenour of our latter-day habits. Probably you, gentle reader, and *you*, and *you*, though very truly women of the world of fashion, have your cherished bit of nineteenth-century sentiment finding its expression in your membership of the " Anti-Fly-Catching Society," or the " Association for Boycotting Butchers."

Not infrequently we wonder whether some day a tender-hearted enthusiastic horticulturist will not found a " Society for the Prevention of Cruelty to Plants." If that day comes, heavy indictments will be proved against those gentlewomen who seem to fill their rooms with plants only to let the poor things die. They will form a very large majority of the ranks of the guilty ; for it is seldom, indeed, that plants which tell tales of neglect or ill-treatment are seen in a cottage. The poor woman does not set up a shelf full of fuchsias and geraniums unless she is sufficiently fond of them not to grudge —amongst all her cares—some time and trouble spent upon them.

The mistress of an artistic drawing-room must have palms, ferns, lilies in pots, etc., to carry out the desired effect ; but unless the gardener has daily access to his exiles, or by good luck the footman or parlourmaid has a sneaking kindness for them, it requires but a few days or a week to deprive the plants of their ornamental character, and turn them into objects of pity, rather than of admiration.

Where an experienced gardener may, in a country house, have his own way with the plants, there is no need for any one else to be anxious. But we must lay down one golden rule for the châtelaine whose gardener is single-handed or inexperienced, and so is fain to let her take her own way with the plants brought into the house for decorative purposes.

The rule is this : Do not keep any plant very long in the house, and never after it shows the smallest signs of deterioration. Let it go back to the more healthful atmosphere of the conservatory, or hot-house, if possible, *before* damage is done. By observing this rule you may enjoy the beauty of the choicest plants, when just in their prime, without any evil results.

Ladies are often heard complaining that the gardener objects so much to their having certain plants brought indoors when in full bloom, although they then form such desirable objects for the centre of the breakfast-table, the stand in the hall, or elsewhere. " Just as if it would hurt them, and as if we might not just as well have the enjoyment of them while they are out." But the gardener remembers only too well how a fine lot of gloxinias languished on that stand in the hall last year, and a beautiful *Crinum* was set in a dark corner of the drawing-room till all its leaves turned yellow and sere, so that the bulb was damaged; and how the azaleas, which were the pride of his heart, were allowed to go dry, and so lost their old leaves, and the young flowering shoots had no chance to form.

A temporary check of this sort may seem of small importance to the amateur, but will, in many cases (as in that of the azaleas), affect the well-being of the plant for at least a year to come.

Other rules to be observed are, to keep the plants clear of draughts, to wash the leaves when dust settles on them, to water whenever required, but not *unless* required, and not to leave the pots standing in water.

In summer, and in clean, fresh country air washing may not be necessary during the whole time a plant is within doors, if only the housemaid be instructed to remove it when she is sweeping the room. But in London, so dusty in summer, so sooty and smoky in winter, washing twice a week should be resorted to with palms and foliage plants. It is the only way of allowing the poor things a chance to breathe, and the length of time they will continue to look well and flourishing under these conditions is really surprising.

Another plea to be put forward for room-plants is, to let them have as much light as possible; the very spot where the presence of a tall graceful plant is generally considered most desirable is too often dark and gloomy. Surely, under these circumstances, it would not be too much trouble to let the occupant of that remote corner stand in the window at such times as the room is unoccupied. Light is not only necessary for the development of chlorophyll, but for its continued existence; and without its agency many life-functions of the plant cannot be performed.

Two more useful hints to those who try to keep a few pet plants in rooms which, from constant use, gaslight, or other reasons, are very dry and hot. First, it will be found to have a reviving effect if in mild weather the pots are placed outside on the window-ledge at night. Secondly, too rapid evaporation may be checked by filling the spaces between the ornamental pot-cover and the actual flower-pot with powdered charcoal.

To revert for a moment to the sentimental side of the question. Are there not certain plants one should prefer before others for *living* with ? These should be defined as plants with a personality as well as a name—plants with pleasant associations or curious habits. Among those whose names will recall some agreeable memory, the geraniums may be admitted for the sake of Evelyn Hope :—

> " She plucked that piece of geranium flower,
>   Beginning to die, too, in the glass."

" Hush ! " says her love to her, as she lies dead,—

> " ' I will give you this leaf to keep.
>   See, I shut it inside the sweet cold hand.
>   There, that is our secret ! go to sleep ;
>   You will wake, and remember, and understand.' "

It is pleasant, too, to sit at the window, and, watching the marguerites, remember not only the humbler flower, but also all the wonderful women who have borne that name—the lovely wilful *Reine Margot*, with her whimsies in dress, her vivid interest in everything, from a new corset to a witty epigram or dainty sonnet, with her keen delight in life, alike in its serious and more buoyant moods. Then there was the *marguerite des marguerites*, the Pearl of the Reformation, Marguerite de Navarre. The pure and holy St. Margaret of Scotland, and that piteous, bewildered sinner, the Gretchen of Faust, owned the same charming name. It was the tinier flower that Chaucer, Burns, Wordsworth, and many another sang. The "gowan" of Scotland, however, the big ox-eye daisy of the fields, is like our window friend, and it is immortalised in the national song that Scotsmen sing on New Year's Eve all over the world :—

> " We twa hae paidelt in the burn,
> And pu'd the gowans fine ;
> But we've wandered mony a weary foot,
> Sin the days o' Auld Lang Syne."

The chrysanthemum is a good plant to have beside one in the autumn. It is the flower of happiness, and redolent of old Japan.   Perhaps, sometimes, it dreams of its old home,—

> " Clear shine the hills ; the rice-fields round
> Two cranes are circling ; sleepy and slow,
> A blue canal, the lake's blue bound
> Breaks at the bamboo bridge ; and lo !
> Touched with the sundown's spirit and glow,
> I see you turn with flirted fan,
> Against the plum-tree's bloomy snow. . . .
> I loved you once in old Japan ! "

The Japanese cook chrysanthemums for their New Year dinner. One almost feels as if they were cannibals !

Musk and lemon-scented verbena are pleasant and fragrant companions to have beside one ; and the verbena recalls Shelley's exquisite lines :—

> " The light clear element which the isle wears
> Is heavy with the scent of lemon-flowers,
> Which floats like mist laden with unseen showers,
> And falls upon the eyelids like faint sleep."

But we need not pursue this suggestion further. Every gentlewoman who loves flowers will have her chosen companions among them.

The arrangement of outside window-boxes will generally, in the country, be left to the gardener, who will use the flowers of which he has a surplus

12

after "bedding out." The usual geraniums and lobelia combination is much to be deprecated. The colours savour too much of cross-stitch, and show to more advantage in mediæval illumination.

Ladies should exercise their wits in trying to plant their boxes in a more original way. Our favourite marguerites and musk often die because it is overlooked that both are *very* thirsty plants, and a very thorough soak should be given in the evening of every dry summer day if the boxes are to look well. After all, the simplest arrangements are the prettiest ; as, for instance, those adopted year after year at a quaint old Manor House, standing off the street of a country town in the Midlands. The house front is covered with ivy right up to the roof, and every one of the many windows has a box, in which are grown simply sweet peas. Festoons of thick wire, along which the peas are loosely trained, pass from one box to another. The effect of the dainty pink blossoms, forming careless loops against a dense background of green, is remarkably good. Scarlet tropæolums used in the same way would be more distinctive, and would last longer.

Again, people are too apt to think that when there is room for anything beyond the regulation row of evergreens for winter, crocuses or hyacinths are the only possible spring flowers. In reality, there are several bulbs available, much less common than those named above; and auriculas or hybrid primroses are most charming for such a purpose.

In choosing evergreens for town window-boxes it should be remembered that broad, smooth-leaved plants are the only ones to be depended on, because when rain falls their surfaces offer some expanse to be washed by it, and some, at least, of their pores, become cleared of the clogging soot. Small or hairy leaves may not get washed at all, and as by degrees all their pores are closed, breathing becomes impossible, the plant grows sickly, and eventually dies.

A point insufficiently studied is, that plants and boxes should be in keeping and in harmony with the house. Scarlet geraniums in blue tiled boxes, against a raw red brick house, form a ghastly spectre of the past that haunts us still. The uninteresting stucco of the average town house is so overpoweringly neutral that all that is brightest should be used to embellish it. A Queen Anne brick house should, however, be allowed to drape itself in greenery as soon as possible, if for no other reason than to provide a harmonious setting to the window decorations.

Before leaving this portion of our subject it may perhaps be interesting to look back to the first attempts at window gardening.

In 1734 Sir Thomas More describes, as a pleasing novelty, " The art of raising flowers without trouble: to blow in full perfection in the depth of winter in bed-chamber, closet, and dining-room."

First of all, he describes how he had "four basons " to be made at the " red ware pot-house," and afterwards painted into an imitation of the "blew ware." They were eighteen inches in diameter, and one foot deep, and were fitted with iron rings, whereby they might be fastened to the window, two inside and two outside of it. In the centre of each of these he placed a Crown Imperial, and encircled it with a wreath of tulips. Outside the tulips, " double white and blew hyacinths " alternated with " anemonys." Yellow and white polyanthus narcissus formed the next ring, and " double daffs " held golden guard outside these. Crocuses marshalled themselves lower down, and snowdrops fringed the edge of the basin ; whilst a couple of fritillarias and a couple of hepaticas interpolated a little variety. Thus, as the astute gardener will perceive, the flowers of the outermost ring bloomed first, and so the succession worked inwards, until at last the Crown Imperial, a stately and dignified survivor, was left blooming alone. Sir Thomas records that in his window, " in the passage against the steeple of St. Bride's Church," looking eastward, snowdrops were flowering outside the. window and crocuses inside on Twelfth Day. The success of his spring " basons " encouraged him to try summer ones; and in these he had " double African, marvel of Peru, capsicum, cockscomb, amaranthus, orange mint," etc.

Nor did Sir Thomas's experiments cease here. It occurred to him that he would try growing plants in water only, thinking it would be " a much neater and cleanlier way, and might be more acceptable to the curious of the fair sex, who must be highly pleased to see a garden growing, and exposing all the beauties of its spring flowers, with the most delicious perfumes thereof, in their chambers and parlours." And this is how he set about it. He purchased some dozens of flint tumblers " of the Germans, who cut them prettily and sell them cheap," also some glass basins, broad at the top, narrower at the base. In these he fitted pieces of cork half an inch thick,—the diminishing size of both basins and tumblers preventing these from sinking. In these corks he cut holes to suit the

size of his roots, and then in the tumblers he set
narcissi, hyacinths, tulips, daffodils, jonquils, etc.,
and in the basins crocuses. His narcissus was in
bloom before Christmas, and he recommends all
" curious ladies " to follow his example. He also
dwells with delight on the fact that artificial heat
was as good as natural heat for the growth of plants.
He grew " young sallets " in his chamber by means
of corks and water. Even the cruel waste of bulbs
caused by his aquatic experiments he considers
quite atoned for by the " surprizing diversion "
they afforded him.

CHAPTER XVI.

THE STILL-ROOM.

"ONCE upon a time," as the children say, the still-room was as necessary a part of the house as the kitchen, and all well-educated gentle-women knew how to distil waters and prepare simple herbal remedies for the smaller ailments of life. In the charming household of Sir Thomas More (of which "Sweet Meg" Roper was the light and life) this ancient art was diligently practised.

Now, in spite of their improved culture, and the more important position occupied by women in the world, this simple art seems almost lost amongst them. One of the great drawbacks to the advance of general knowledge is that it leaves less room for individual ingenuity and information. We have quantities of "specialists" in every line, and there is a tendency to trust blindly to them, and take no trouble to find out things for ourselves. The individual is indeed "withering" with a vengeance, and soon will be as extinct a species as the mega-therium.

But before it is too late, we must make a protest and put in a plea for the renewed cult of indivi-duality, especially in our own department.

There is nothing in which the individual is more necessary than in the culture and care of flowers; for each flower is an individual, and needs individual comprehension and attention.

Why is it that, in the rage for occupation nowa-days, no one has thought of reviving the still-room ? Nowhere could the gentlewoman find a more dainty occupation. Compared with it, painting, pokering, and gardening itself are rude, toilsome, and dirty occupations. Who, that has looked into a labora-tory, can fail to be charmed with the graceful glass apparatus—slender test tubes, measuring-glasses, and so on ; the balances, all of polished wood and shining brass, with their tiny weighing trays and infinitesimal weights. Chemistry is indeed much more womanly work than many people think. It requires the most delicate and careful hands, the gentlest fingering, and the nicest cleanliness. To-day, the apparatus employed is infinitely more delicate and attractive than that with which our ancestresses had to work. All utensils possible are

made of the finest Bohemian glass ; for, as every
good housewife knows, the finest, thinnest glass is
the strongest. The fair chemist has no longer to
soil her fingers by attending to a coal fire, but heats
her liquids in thin glass flasks over a Bunsen gas-
burner.

Before proceeding further, it may perhaps be as
well to explain for those who do not understand it
the gentle art of distillation. The object of distillation is to separate a desired liquid from undesired
impurities—whether liquid or solid—by the aid
of heat. Take lavender, for instance. You wish to
steal and retain its perfume before the flower dies
and withers, and this you can best do by means of
distillation. In distillation there are three stages :
first, you transform your original substance—which
if not liquid must be made liquid by means of heat
—into a vapour ; next, you turn your vapour, by
means of a cooling apparatus, called a condenser,
into a liquid again ; and, lastly, you catch and
retain the new liquid in a vessel called a receiver.
At first the advantages of this process are not
apparent. What, asks the indignant reader, is
the use of turning a liquid into a vapour, and then
back to a liquid again ? But the principle of the
art is this—the perfume that we covet is carried

by the flowers in tiny, almost invisible sacs or bags. Steep these flowers in water, apply heat to them, and the perfume, being a volatile (the word explains itself) oil, escapes from the control of the flower, and takes the form of a vapour. The other portions of the plant, being less volatile, remain behind, and we have what we want, the perfume

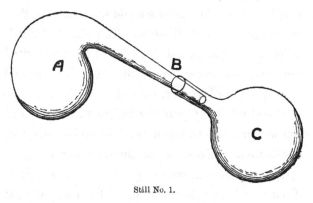

Still No. 1.

separated from its less essential and interesting companions. The vapour is passed at once into a condenser, kept cool by a stream of water, where it regains its liquid form, and trickles thence into the receiver, to be bottled off for use.

We shall now proceed to describe the apparatus the gentlewoman will require to have in her still-room. First of all comes the still ; in its simplest form a glass retort and receiver, such as this. You put the herbs first into A. The vapour passes into B, and is there condensed by means of moist blotting-paper, on which cold water drops, so that by the time it reaches C it is a liquid. Properly speaking,

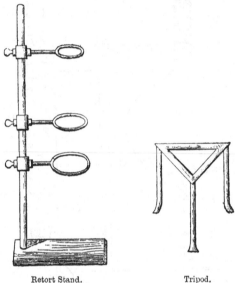

Retort Stand.   Tripod.

A is called a retort, and should be placed either in a retort stand or on an iron tripod beneath which must be placed a Bunsen's gas-burner.

Retorts are usually employed for distilling small quantities, although they can be had in sizes up to the holding capacity of a gallon.

Another simple still is this. A is the retort, B is what is called a "worm-condenser," C the receiver,

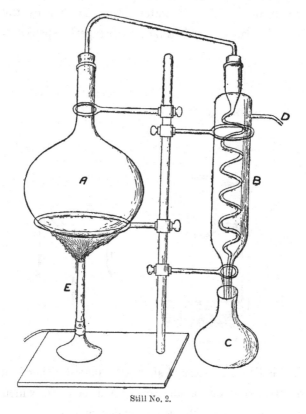

Still No. 2.

D is an escape pipe for the hot water, and E the Bunsen's burner. In the worm-condenser the vapour passes through the worm-like tube in the centre, whilst cold water trickles round it from a reservoir above. The vapour, of course, heats the water ; but the hotter water, being lighter than the cold, rises to the surface and escapes by D, whilst

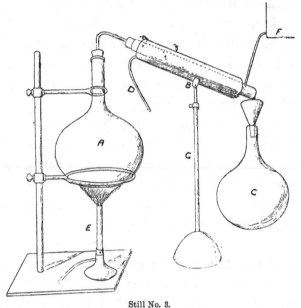

Still No. 3.

fresh cold water comes from the reservoir to take its place.

One more still, and we have done. This, it will be at once perceived, is on the same principle as No. 2, but has a Liebig's instead of a worm-condenser. A, B, C, D, E, are the same as in the former illustration, whilst F is the cold water reservoir and E a support for the condenser.

The still, an ordinary balance for weighing with, and a graduated measuring glass of 12 oz. capacity, are about all that the lady of the still requires.

Having thus described the apparatus of the still-room, we shall now suggest a few such simple things as a gentlewoman may do there. Rose water, elder-flower water, acacia-flower water, bean water, are all made after the same recipe.

It should be explained that there are single and double waters. This is the single water: Put 1 lb. of flowers and 1 qt. (40 ozs.) of water in the retort and distil off 16 ozs. To obtain the double water, return the single water to the retort with 6 ozs. of fresh flowers, and distil off.

Lavender water is a thing that every one will want to distil. To make good lavender water it is best to use spirit as well as water. Put ½ lb. of lavender flowers (free from stalks), 1 qt. of rectified spirit (60 over proof), and 2 qts. water into the retort, and distil off one quart.

We should mention that receivers with measures already marked on them can be obtained. A simpler and less expensive plan, however, is to have a series of receivers, each for a special quantity. The stills can easily be had from any of the dealers in chemical apparatus.

There is one lady not far from London who carries on a very successful business as a lavender and mint grower and distiller. She inherited the business from her grandfather, and she has had the enterprise and skill to maintain it.

Lavender water is not always lavender water pure and simple. To obtain a good perfume others are often added, such as rose water, tincture of orris, oil of bergamot, essence of ambergris, oil of cloves, musk, and oil of rosemary. The gentlewoman, however, will no doubt be content to make her lavender water pure. There are many perfumes that would lose their delicacy by being distilled. The methods usually employed to retain these are *enfleurage* and *maceration*.

Many flowers lose their perfume when they die, and, therefore, it is necessary to press the living flower upon some oleaginous substance. The greasy matter at once absorbs and retains the perfume. The grease is afterwards put into spirit, and the perfume having a greater affinity for spirit than for fat, at once abandons its first love, and gaily unites itself to its second. The odourless grease is removed, and the perfume is ready. A simple experiment will prove this. Grease two plates thickly with unsalted butter. Lay fragrant clematis on one plate, and cover with the other, and in twenty-four hours you will have scented butter! This is called *enfleurage*, and is used for roses, orange-flowers, acacias, violets, tuberoses, and jonquils. There is yet another method employed at times for jasmine. Layers of cotton-wool soaked in oil of ben are prepared, and layers of jasmine flowers alternated with them. This the gentlewoman might do in her still-room.

The leaves of herbs such as sage, marjoram, savory, dill, rosemary, thyme, and peppermint, may also be distilled either with water alone, or with a mixture of spirit and water. In this form the herbs are more suitable for delicate culinary operations than when dried and rubbed, as they usually are. Peppermint and dill are also useful medicinally; and the latter is almost indispensable in a house where there are children.

Horehound is a simple remedy for colds that might be prepared in the still-room. Make a decoction of the dried herb with the seed, or the juice of the green herb, and mix it with honey before giving it to your patient. The decoction of the dried herb may be made by infusion, but the juice of the green herb should be pressed out when it is young and tender by bruising it in a stone mortar with a wooden pestle. Put into a canvas bag, press it hard, then take the juice and clarify it. You clarify it by putting it on the fire in a stone vessel, and skimming the impurities as they rise to the surface. When cold, put it in a glass, and pour a little oil over it to keep the air out, and prevent putrefaction.

The art of distillation is a very ancient one. It originated, like everything else, in Egypt, and the Greeks and Romans took it thence. Neroli, a preparation of orange-flowers, was made in old Roman days, and called after the Emperor Nero.

Many pleasant thoughts and picturesque memories circle round the still. In Switzerland the peasants climb the hills to where lavender, rosemary, and thyme grow in abundance, and distil the perfumes on the spot. Here is a suggestion for a novel and delectable summer trip. Think of camping out on the hillside amidst the fragrant flowers, and wiling away the long and dreamy summer afternoon by filling the retort, and watching the leisurely distillation of the herbs. It would be rather risky to use a glass still for this. The copper or tin still, such as our ancestresses used, would suit the

13

purpose better. As they do not supply gas on the hillside, the charm of gathering a brushwood fire beneath your retort would be added to the others. If you had a minute left to sit and dream in, you could not but remember the "bank whereon the wild thyme blows" dear to every lover of Shakespeare.

> "Pun-provoking Thyme"

Shenstone calls it ; and Tennyson talks of the

> "Thymy plots of Paradise."

As for rosemary, the old practice of placing it on graves makes it a somewhat melancholy herb—

> "Rosemary is for remembrance
> Between us daie and night ;
> Wishing that I might alwaies have
> You present in my sight."

Ophelia and Perdita both loved and remembered rosemary.

A herb-garden is a natural adjunct to a still-room, and nothing could be more delightful. Herbs are mostly quaint, old-fashioned plants. Lavender has a frosty, subdued, old-maidy look, and rue, an orderly grace about it. Betony is straggly and uncouth, but fennel is one of the prettiest and most graceful plants that grow. Mint is not attractive, and has a forward way with it, which, but for the thought of the early lamb,

might . cause us to dispense with its presence.
Sage should be admitted for the sake of Charles
Lamb. He permitted a " dash of mild sage" to
accompany his " child-pig" ("consider he is a weak-
ling—a flower ") to table. " Sweet marjoram" who
could resist ? It is " in shepherd's posie found "—
there is music in the very name. They all grew—
and many more—in the garden of Shenstone's
delightful old school-dame. Spenser has a very
quaint list : —

> " The wholesome Sage and Lavender still grey,
>   Rank-smelling Rue, and Cummin good for eyes,
>   Sharp Isope, good for green wounds' remedies,
> Fair Marigolds and bees-alluring Thime,
> Sweet Marjoram and Daysies, decking prime.

> " Coole Violets and Orpine, growing still,
>   Embattled Balme and cheerful Galingale ;
>   Fresh Costmarie and breathfull Camomile,
>   Dull Poppy, and drinkquickening Setuall,
>   Veine-healing Verve, and head-purging Dill,
>   Sound Savorie, and Bazil hartie-hale,
> Fat Colewort and comforting Perseline,
> Cold Lettuce and refreshing Rosemarine."

In this brief chapter we have not touched on a
hundredth part of all that the gentlewoman might
do in her still-room. That cool haunt has endless
possibilities, and distillation could readily be practised without any special chemical education or

training. A lesson or two in the management of the still from a practical chemist is all that would be required.

The glass apparatus is of course apt to break, but with ordinary care in regulating the heat this need seldom occur ; and it is certainly much easier to keep clean, and much pleasanter to work with than. the copper still.

CHAPTER XVII.

HOME AND TABLE DECORATION.

A GREAT deal that was said in the last chapter naturally applies to this also. Growing plants are largely depended on for decoration of the dinner table, rooms, corridors, etc. Perhaps some day ladies will be induced to believe that pomp and circumstance of surroundings will not make an unhealthy plant agreeable to look upon ; a diseased and shrivelling palm, though reared upon a gilded column, is none the less sickly or faded. If they cannot be at the trouble of attending to their plants, or, for economy's sake, cannot employ some capable person to do so, they really had better eschew their use, or content themselves with those artificial dracænas and foliage begonias which one sees on the refreshment counters at railway stations. These, at least, want no watering, and suffer not from red spider, thrips, or scale ; the housemaid, with her duster, will be quite competent to deal with them.

"As if I would have such abominations in my house ! " cries a chorus of indignant voices. "Had we not better substitute, also, clean-washed anti-macassars of Nottingham lace for a Turkish embroidered chair-back, which may perhaps be soiled a little ? "

Certainly, if you choose, and think the cases parallel ; but as most persons of taste would prefer to drive a sound pony, well-groomed, in an unpretentious cart, rather than a pair of screws in a dilapidated brougham, so good taste (when directed to the subject) will perceive that a fresh and healthy mignonette or two, cheap and easily renewed, are preferable to a regiment of neglected moribund palms.

Again, unless a hostess is able and willing to renew very frequently the little ferns, etc., which it is just now fashionable to dot about rooms and tables, she must be content to leave them in ordinary flower-pots, or, at any rate, in something porous. In the pretty fancy bowls, whether metal or china, the plants can only exist a very short time, as there is no drainage, and without this the roots cannot possibly thrive.

Bold effects are always artistically preferable to the multiplication of small and trivial ornament ;

and therefore a few large plants, judiciously placed, will always be more desirable than quantities of miniature pots stuck about here, there, and everywhere. So it is greatly to be wished that every mistress of a house will think it worth her while to acquire the small amount of knowledge requisite to keep these specimen plants in order, when it happens she has not the gardener's experience to trust to in the matter.

The use of tall plants is only to be deplored on a dinner-table, for, however effective they may be as a spectacle, when viewed from the door of the room, for instance, they become to those seated at table mere obstacles to sociability. The decorator may spread her plants as much as she likes, but let her keep them low.

As to the floral decoration of dinner-tables, the great thing to be aimed at is to make them uncommon, if possible ; and this does not necessarily mean that rare and costly flowers must be used : often the greatest successes are obtained with the simplest materials. Every one knows, for instance, that common daffodils, tastefully arranged, are not easily to be matched for beauty as a decoration. One of the prettiest mixtures we remember to have seen was in April, at a very small and modest country house ; in fact, it was only a farmhouse, a little enlarged and beautified to meet the demands of a young couple who began their married life there. The sunny garden, sloping south and fenced with yew-hedges, and a single greenhouse, were all that the young hostess had at her command : the garden supplied her with a quantity of single white primroses (grown from seed), and with these she had filled low bowls of clear glass, placing gracefully in each a few sprays of deep pink geranium from the greenhouse. This combination lighted up beautifully, and the charming result has remained a pleasant memory.

In mid-September a beautiful effect can be produced by placing clusters of scarlet geranium among trails of the wild clematis, known as " Old Man's Beard." When the seeds first begin to form, before the white, fluffy stage is reached, there is a soft, silvery lustre on the carpels, which tells extremely well by lamplight.

What can be more charming than small wild roses, mixed with forget-me-nots, in June ? Or again, in October, the bronzed leaves and deep red berries of the guelder roses make a perfect tone picture when combined with the soft French-grey of the wild scabious. But it is only defeating one's

object to continue giving suggestions, for if they were
to be adopted by those readers who cannot devise
novelties for themselves, then the very monotony
would be encouraged which it is so desirable to
prevent.

It is not much credit to have one's rooms luxuri-
antly furnished with flowers if one has at one's
back acres of splendid gardens and ranges of glass-
houses.   Yet where this is the case, how very often
stiff and tasteless decorations prevail, because they
are left to the gardener and his assistants, who desire
only to show off the wealth of their resources.

Necessity is the mother of invention, in the world
of floral decoration as in that of Science.   A mental
picture kept side by side with that of the white
primroses and pink geraniums, is of a gipsy basket
filled with brown bracken and yellow sprays of
Jasminum nudiflorum.   This was the January
decoration of a rectory drawing-room, where there
was no glass to supply cut flowers during winter.
Often, too, one sees beautiful subjects unexpectedly,
and can utilise them immediately ; as once when
paying a visit towards the end of August we were
asked to arrange flowers for an afternoon party.
The drawing-room was large, light, but sunless ;
and though crowded with valuable objects, had a

want of warm colour, enhanced by cold, grey walls.
Wandering out rather hopelessly to see what the
garden afforded, we lighted on some clumps of
peonies, with their great spreading leaves turned
purple by the breath of coming autumn. These,
in brackets and tall jars, used as a background for
spikes of gladioli—red and salmon-coloured—gave
the requisite brightness and warmth at once.

This brings us to the subject of receptacles for
cut flowers. When rooms of any size have to be
decorated, a few bold arrangements in large jars
or bowls are infinitely more appropriate than little
insignificant globes, tubes, and vases, of which an
immense number are wanted to "tell" in a large
space, while they are utterly unworthy to display
flowers of the size which may be ventured upon in
such cases.

In these days, when a revived fancy for the
display of plate has caused every old piece to be
dragged out for use or show, it not unfrequently
happens that flowers suffer by being thrust into
some squat, clumsy silver basin or cup, originally
intended for ale, or maybe for caudle !

There is a little Horticultural Show we know of,
in a certain little town, where year after year the
same prizes are taken by the same people. For

many seasons back the gardener from one establish-
ment has always been first for a "bouquet of cut
flowers"; not that there is the least particle of
taste displayed in his arrangements, but because
they are contained in a vase with a silver pedestal.
This bias, on the part of the judges, used to be a
cause of perennial amusement; but time has brought
its revenge, and Fashion to-day justifies them of
their choice.   No amount of ugliness can disqualify
a modern bouquet, if only it be in a silver bowl—
so, at least, says the silver-maniac.

When this phase has passed away, it will be
realised once more that beauty of form and firmness
of base are the positive virtues of a flower vase ;
while its negative ones consist in absence of strong
colour, and of inaccessible corners which cannot be
washed clean.

A fantastic shape is not really beautiful for such
a purpose ; the simpler the lines the better, so long
as they are flowing and graceful ; and as for colour,
if any one predominates, then the vase should be
used for flowers of that tint only, or those com-
plementary to it.   Soft dull greens, brown, and
creamy whites, are the most useful " all-round "
colours for china or pottery vases.   And in glass
there is nothing like the olive-green of Bohemia.

Whether one's flower glasses are costly or not, it is best to have several sets for the dinner table, and a few surplus ones, admitting of an occasional change, in the other parts of the house. If one always uses just the same vases it is difficult to avoid monotony of arrangement, let one's materials be ever so various. The very variety, in fact, calls for differently shaped receptacles, for it is a barbarity to snip off short some flower which grows on a long stalk, and to thrust its diminished head into a low bowl or tiny globe. Picture a tall trumpet-shaped vase, or upright tube, placed where a dark, rich background throws up its burden—say of four or five tritoma spikes, with their long ribbon-like leaves, or pure white callas and their handsome foliage, or perhaps a great branching head of *Lilium Auratum*, breathing out spicy odours. Compare their effect, so treated, with that produced if you cut them off short, and crammed the blossoms in a flat dish. To good taste it should be as impossible to do this as to arrange soft, heavy-headed roses in the tall trumpet-vase.

This chapter must not be closed without some reference to the more formal decorations necessary on the occasion of balls or evening parties. Suggestions for these are not offered, because a woman of taste can always devise something pretty and suitable out of the material at command, while those who have not taste should let it alone, and employ some poorer · lady who has this valuable gift.

There are some few fancies of fashionable florists which look delightful when the room is empty, swept, and garnished ; but which do not certainly tend to the comfort of the guests when assembled.

It is not pleasant, when nearly carried off your feet on a crowded staircase, to grasp at the balustrade, and find you are crushing ropes of flowers, without gaining any support yourself ; or, when a room is denuded of all its furniture except chairs, piano, and console tables, to discover that the last, together with the mantelpiece, are completely banked with moss and flowers, so that there is no coign of vantage whereon a weary arm may rest, or even a fan be laid down for a moment.

Groups of plants, too, at the head of the stairs, or incidentally occurring in the corridor, are traps for the unwary. A nervous or short-sighted person is sure to blunder into them ; or the little outside pots get swept away by some dowager's heavy train, and roll about, or trip some one up.

Decorate, decorate, decorate, by all means, if you can afford it—you are promoting a deserving industry ; but don't let the plants and flowers be a source of discomfort.

The fact is, that for crowded rooms we want to move our decoration higher up ; bestowing the greater part of it on the top of the doorways, the upper walls, and the ceilings.

CHAPTER XVIII.

GARDENING AS A PROFESSION.

THE parent ideas of lady-gardening are (1) the necessity, through over-population, and the surplus number of unwedded women, of providing methods by which a woman may earn her living ; and (2) the love of nature called to fresh life early in the century by Wordsworth's poems, and fostered by the charming writings of such true nature-lovers as Richard Jefferies, and by the frequent " nature " articles in the leading newspapers.

The lady-gardener is, emphatically, a modern product ; but as the seed of every modern idea was laid far back in the centuries, so was this, and we seem to see it in the pleasant old still-room practices already referred to. Do you think the country dames who distilled their own lavender, for instance, were content to leave the growth of the herb entirely to others ? No; we are sure these gentlewomen paid frequent visits to their gardens, noted the heads of flowers, and the increase and decrease of luxuriance in the plants; and we have no doubt they took an interest and a share in the nurture of their still-room herbs and flowers. The still-room has fallen into abeyance—although we hope it may soon be revived—but the lady-gardener is daily becoming more *en evidence*.

It may be said that in a book of "Gardening for Gentlewomen" this chapter need have no place; but it is exactly because it is for gentlewomen that we have inserted it.

The papers are full of the distressed gentlewoman, and her only resources so far seem to be "art" needlework (in which there is but a pinch of "art" to a vast quantity of needle) and "decorative" painting, by means of which one's drawing-rooms have accumulated such a vast amount of ill-painted white wood, on whose sticky surface the dust collects with fatal rapidity, thus necessitating their being speedily replaced by new monstrosities. Both these occupations, the refuge of the incompetent, are overstocked. The lady doctor requires more than the average nerve and courage of women; whilst the lady milliner perishes, as a rule, betwixt Scylla and Charybdis. She cannot determine whether to be lady or milliner, and consequently succeeds as neither. There have, of course, been a multitude of other occupations experimented on; but we cannot believe that the Stock Exchange, the balancing of commercial accounts, the practice of law, or the manipulation of drainpipes, are professions that will readily commend themselves to our ideal gentlewoman.

But where is a lady more at home than amongst her flowers ? Where does she seem fairer, more womanly, or more healthful ? This out-door occupation is the very thing wanted for strengthening delicate girls. The children of the upper classes suffer from too little muscular exercise. The child of the working man or of the middle class, where one servant or no servant necessitates the performance of much *real* work by the daughter of the house, has a much better chance in the struggle for existence than the too-carefully served and tended gentlewoman.

Sports have done much ; but there are, we trust, women to whom the feeling of *usefulness* adds a charm to exercise ; and to these the care of the home-garden should be a welcome and enjoyable task.

As for the gentlewomen who have to earn their living,—what more natural occupation than this ?

14

Many of these have been brought up in a pleasant country house, which, after a time, owing to the improvidence or misfortune of their parents, they have to leave in order to gain a livelihood. They have been used, perhaps, to "potter" about the garden ; they know something about the flowers and their needs ; and by studying so as to transform their desultory into practical knowledge, they may easily qualify themselves for starting as lady-gardeners, and so continue the pleasant country life they have been accustomed to. The only difference is, that from being drones they have become busy bees in the hive of life ; and surely at this later day there is no need to preach the gospel of work.

The qualifications of the lady-gardener are energy, patience, and perseverance. Energy and a certain amount of patience many women have, but perseverance in hard and often disheartening work is not the attribute of all, and no one must expect to succeed as a gardener without it. The lady-gardener must be always there. Flowers are like children, beautiful, irresponsible, helpless things ; and we may say that the good gardener will make a good mother, and the good mother—after her children are grown up—the best gardener. When roses and

babies compete, the roses always come off second
best. Still, like the children, the flowers naturally
make for health, and, given favourable conditions
and tender care, will reward their guardians by
their beauty and luxuriance.

Gardening necessitates early rising and late
attendance at the greenhouse stoves. The actual
rough work, such as digging, etc., will always be
done by men, but there is much toilsome stooping
and bending to be endured. Practice soon renders
this easier, and makes the muscles more supple, and
the staying power greater. Fresh air is a magnificent stimulant. The old fable of the warrior who
renewed his strength every time he came in contact
with mother earth, is, like many of the Greek
stories, but a picturesque expression of one of the
fundamental laws of nature.

Profitable gardening, like most other things nowadays, is worked by specialities ; and a woman who
takes up this occupation seriously should turn her
attention to the growth of two or three suitable
crops, each in perfection. For choice cut flowers
there is always a good market ; and we know one
gentleman grower who makes a good profit by
keeping to just a few kinds of flowers, such as
lilies of the valley, gardenias, etc. *The* reason of

this is clear to all—the amount, and therefore the cost, of work required is so much reduced and simplified. A great deal lies in growing the *right* flowers, some being more fashionable, and therefore more saleable, than others ; and it is Fashion that rules the markets, so that it must not be ignored by the would-be lady-grower.

There are three distinct lines into which the work falls ; all have been adopted by women, though the first seldom, and hardly ever, with success.

The first is that of nursery-gardening, growing trees, shrubs, plants, bulbs, and seeds for sale ; the second as a market-grower, supplying vegetables, flowers, or fruits in large quantities for Covent Garden or other markets ; the third is the retail-grower, who cultivates numerous small general crops for private sale. The possession of that business faculty which is *born* in some people, but not in others, is necessary for the first two ways.

Fortunately, Napoleon's saying that we are a nation of shopkeepers is mostly true; the commercial feeling is more strongly distributed in England than in many countries. It is less common in women than in men ; but many women have it, and those who have it do well to make use of it. If they do not require to earn their living, let them give their services to one of the numerous charities to whom such aid is valuable. We all know what the "managing" woman is when confined to "the trivial round, the common task."

The third method suggested is, of course, the pleasantest for gentlewomen who have been brought up in retirement, and do not feel equal to chaffering with buyers who want to pay too little.

There are many ladies throughout the country making a moderate competence by gardening, and there are now ample opportunities for educated women to obtain the training which is necessary for success as professional gardeners. They are too apt to fancy their untrained services valuable, or that teaching should be *gratis*. To meet the demand for training that is now arising, a Ladies' Branch of the Horticultural College at Swanley, Kent, was opened in June 1891. For the sum of £70 to £80 per annum the female students are boarded in a bright and comfortable home, close to the college, in the grounds and lecture-rooms of which they pursue their studies. The course includes botany, chemistry, zoology, physics, building, construction, and book-keeping. To some minds much of this appears unnecessary; but it should be remembered that in this artificial age the gardener is always trying to forestall Nature, and only Science can help him (or her) to achieve this.

Many of the private lady-growers take pupils ; but while their charges are higher than those of the college referred to, theoretic teaching is not given, and practical experience can only be obtained in the particular line which the teachers have adopted.

We do not conceal that many unfit persons have tried gardening, and the result has been unfortunate. These have mostly been women who have not received thorough training, and have rashly attempted to be " professional " without understanding all the word implies.   Others have failed through a foolish pride, which prevents them, like the milliners aforesaid, from regarding the matter purely as one of business.

Floral decoration and window gardening are charming occupations for ladies to take up ; and surely there are many who would prefer to have a softly footed and shod lady coming in and out of their houses to tend their window-boxes. The average man has a singular talent for displacing things, and no idea of replacing them. Besides, women always take a more kindly interest in and care of any carpets and furniture that may come in the way than men.

A pretty and pathetic task that might fitly fall to the charge of the lady-gardener is the care of graves. There is a suggestion of immortality about flowers which makes it seem especially suitable that the gloom of a tombstone should be lightened by their presence. Flowers, after their winter sleep, awake fresh and bright in spring, and the thought brings consolation to those left behind. We well remember our first visit to George Eliot's grave on the heights above London. The city lay misty and dreamlike below; other graves around were bright with flowers, but the space in front of her sombre granite tombstone was plain and flowerless—and this was in August! The words on the stone (her own)—

> " Of those immortal dead who live again
> In minds made better by their presence "—

were noble and stirring in their way; yet to the ordinary woman a single rose blooming above her would have suggested pleasanter thoughts and brighter hopes.

How different the lowly grave of Rhoda Garrett, under the chancel wall of a Sussex village church ! On and around it, in profuse but " careless " order, are all kinds of sweet old-world flowers, provided so that some shall be in blossom every month of the year.

Richard Jefferies maintained that no nations loved Nature better than the English, and he says : "Those who really wish their gardens, or grounds, or any place, beautiful must get that greatest of geniuses, Nature, to help them, and to give their artist freedom to paint to fancy, for it is Nature's imagination which delights so. . . . In Italy, the art-country, they cut down the ilex-trees, and get the surveyor's pupil, with straight edge and rules, to put it right and square for them. Our over-educated and well-to-do people set iron railings round about their blank pleasure-grounds, which the potato field laughs at in bright poppies."

Richard Jefferies has pleaded well for Nature— and not in vain. Before we loved her from afar, took a general interest in sunsets, hills, and other large things, we could not help seeing; but he has brought her close to us, and familiarised us with her most trivial facts and fancies. In a charming paper on " Wild Flowers " he describes the growth of his love of flowers. First he was content to see and gather them; then arose a desire to know their names. This led to discriminating between them, to noting their different individualities. The gardener goes further, and seeks to care for them as well.

Wordsworth's " Confession of Faith " already referred to in this book,—

> " 'Tis my faith that every flower
> Enjoys the air it breathes "—

is very dear to us, and should be present in the mind of every true gardener.

To Wordsworth, indeed, the flowers told their deepest secrets.   When in their presence he forgets to be didactic, and remembering only to be happy, bubbles over with song.   For the little celandine and the daisy he has a myriad of tender thoughts and an abundance of dainty imagery ; his daffodils are a delight for ever, and he has a charming sweetbriar standing on tiptoe to see over the honeysuckle's shoulder.

> " In the woods
> A love enthusiast, and among the fields,
> Itinerant in this labour, he had passed
> The better portion of his time ; and there
> Spontaneously had his affections thriven .
> Amid the bounties of the year, the peace
> And liberty of nature.   These he kept
> In solitude and solitary thought,
> His mind in a just equipoise of love ;
> Serene it was, unclouded by the care
> Of ordinary life."

It is this "peace and liberty," this serenity of mind, that a simple open-air occupation such as gardening fosters ; and it is just these qualities that are yearly becoming more uncommon amongst women. The present-day woman is all cosmetics, all whalebone, all nerves. She has created the demand for the nerve-doctor, for the masseur, for new stimulants such as brain-phosph, etc. Let her leave her toilet-pots, abandon at least a portion of her whalebone, work in the open air, and in six months her nerves will have ceased to trouble her.

It is the privilege of Nature

> "To lead
> From joy to joy ; she can so inform
> The mind that is within us, so impress
> With quietness and beauty, . . .
> That neither evil tongues . . .
> Nor greetings where no kindness is, nor all
> The dreary intercourse of daily life,
> Shall e'er prevail against us."

Printed by Hazell, Watson, & Viney, Ld., London and Aylesbury.

Printed in the United States
By Bookmasters